REGULATION THEORY AND THE BRITISH STATE

Regulation Theory and the British State

The case of the Urban Development Corporation

MO O'TOOLE
Department of Politics
Newcastle University

Avebury
Aldershot • Brookfield USA • Hong Kong • Singapore • Sydney

Published by
Avebury
Ashgate Publishing Limited
Gower House
Croft Road
Aldershot
Hants GU11 3HR
England

Ashgate Publishing Company
Old Post Road
Brookfield
Vermont 05036
USA

British Library Cataloguing in Publication Data

O'Toole, Mo
Regulation theory and the British state: the case of the Urban Development Corporation
1. City planning – Great Britain
2. Cities and towns – Growth – Great Britain
3. Urban policy – Great Britain
4. Community development corporations – Great Britain
I. Title
351.8'18'0941

Library of Congress Catalog Card Number: 96-83250

ISBN 1 85972 305 5

Typeset by The Setting Studio, 121 Newbridge Street, Newcastle upon Tyne.

Printed and bound in Great Britain by Ipswich Book Co. Ltd., Ipswich, Suffolk

Contents

Figures

Abbreviations

AES	Alternative Economic Strategy
AMA	Association of Metropolitan Authorities
BDC	Bristol Development Corporation
CIPFA	Chartered Institute of Public Finance and Accounts
CLES	Centre For Local Economic Strategies
CPO	Compulsory Purchase Order
DCC	Dockland Consultative Committee
DT	Department of Transport
EAP	Environmental Area Programme
EFL	External Financing Limit
ERDF	European Regional Development Fund
GEC	General Electric Company
GLC	Greater London Council
INLOGOV	Institute For Local Government Studies
LDDC	London Docklands Development Corporation
LRT	Light Rapid Transport
MDC	Merseyside Development Corporation
MOD	Ministry of Defence
MP	Member of Parliament
MSCC	Manchester Ship Canal Company
NAO	National Audit Office
NCVO	National Council for Voluntary Organisations
NESL	North East Shipbuilders Limited
NLF	National Loans Fund
NTDC	New Town Development Corporation
NTMBC	North Tyneside Metropolitan Borough Council

P&R	Policy and Resources Committee
SDA	Scottish Development Company
SWEB	South West Electricity Board
TPMBC	Trafford Park Metropolitan Borough Council
TPUDC	Trafford Park Urban Development Corporation
TWDC	Tyne & Wear Development Corporation
UDC	Urban Development Corporation
UDP	Unitary Development Plan

Acknowledgements

This book illustrates many of the shortcomings – but I hope also some of the merits of converting a PhD into published material. The PhD itself was long in the making. But I feel that the issues which it explored at the outset still have significant resonance today. Numerous people have assisted me in this project, I am grateful to officers and members at both UDCs and local authorities who took the time to answer my questions. Nick Oatley and Andrew May in Bristol and Alastair Balls and his staff at Tyne & Wear Development Corporation deserve special mention. Amongst my academic colleagues Patsy Healey gave me the confidence to explore my own enthusiasm for theory – she always felt I was onto something. Murray Stewart clarified my thoughts first time round – I hope he will be happier with this attempt. Rachel Lart gave me much more support than an academic colleague should have to. Gerry Stoker pushed me to get my act together and encouraged me to use my own voice. Rod Rhodes along with Martin Harrop subsequently read my work and prompted me to polish up my style. Sarah Golightly was infinitely patient with my eccentric correcting and rewriting and worked with me way beyond the call of duty to get it all finished on time. She was aided in this by Susan Latif – who, as ever, came to the rescue when deadlines loomed. Amy and her friends in Manchester remain in my thoughts – this is still a tribute to them. But perhaps most of all Avram coaxed and corrected from me, something better than I would have originally given. His exasperation at the waywardness and ambitiousness of my work was only countered by his careful assistance in shaping the book. Thanks to you all. I hope you like it.

Mo O'Toole
Newcastle upon Tyne
1 March 1996

Introduction

> ‘Political experience taught the bourgeoisie another lesson that could be used to check the undue radicalism of any urban based political movement: superior control over space provided a powerful weapon in class struggle’
>
> David Harvey (Harvey 1985a).

> ‘The prospect of a long and costly fight with local government and local interests, both conservative and anti conservative, must make it less likely that urban communities will seize, without active encouragement the opportunity for self help which they are now to be given. .....All our experience shows that vested political interests have a great power to delay social progress...Ministers should therefore have enough confidence in the benefits that they are offering to take reserve powers to overcome local obstruction. Legislation should ensure that any freedoms from local bureaucracy which Parliament decides to confer will then be available without restriction to those who wish to take them up’
>
> Michael Heseltine (Heseltine 1990).

Introduction

This book is about the Urban Development Corporations. It attempts to explain their role in reorienting city governance and the recent practices of urban regeneration in the United Kingdom. While it draws on empirical research that was carried out during the 1980's and the early 1990's, its primary task is to place the analysis of these agencies within a theoretical framework that helps us to interpret the transformation which is currently occurring in the state form.

There have been some dramatic changes in state institutions and their policy making processes during the last twenty years (Cloke (ed) 1992). Externally the state's ability to transform public expectations about local economic intervention has been used unsparingly and to some significant effect (Stewart and Stoker 1989). These changes however, including the imposition of flagship policies such as the UDCs, have not gone unchallenged. It is questionable that 'Thatcherism' has been effective on every front, in fact the evidence is that overall, the 'Thatcher Project' of reorganising the state form has been somewhat patchy and only selectively successful.

In order to unpack some of the complexity involved in the implementation of the UDC initiative, I want to do three things in the book. Firstly, I want to place the analysis of state institutions within a regional context. In order to do this I will draw on the work of the Marxist geographer, David Harvey (Harvey 1985). Secondly I want to situate the transformations of the state within the historical context of the capitalist economy, and here I rely heavily on the influence of the Regulation School, particularly the work of Henk Overbeek (Overbeek 1990). Finally I want to add to this analysis, an understanding of the mechanisms of process change within the state and to this end I draw on the work of recent neo marxist policy scientists and analysts (Offe 1984, Benson 1983).

The objective in bringing this somewhat eclectic ensemble together is to illustrate the heterogeneous nature of the state form and to provide a picture of the pace and level of state change, within a nation state that is subject to massive institutional and ideological crisis. I want to explore the struggles that have been, and are are taking place over the nature of the state form and relate them to historical and spatial factors, genuflecting especially towards some of the more recent work that has been carried out on agents reflexivity (Beck et al. 1994, Hay 1995).

The Background to the Book

Urban Development Corporations were created as a result of the 1980 Local Government Planning and Land Act. They were established as targeted single issue urban regeneration agencies operating in inner cities, hence they claimed all the planning and development control responsibilities for their designated boundary. Their job was to reclaim, revalue and repackage derelict brown field sites (Moore and Catchpole 1983).

It has been suggested that they had another objective which became more explicit on the evening of the third consecutive electoral victory for the Conservative Party in 1987. Prime Minister Thatcher, seeing some remaining bastions of 'leftist' power in Labour urban authorities, vowed to take her ideological battle to the inner cities (MacGregor and Pimlott 1990). This interpretation suggests that the UDCs would provide an alternative political and administrative model for the delivery of local planning and economic development policy (Stoker 1988).

UDCs were emblematic of the Thatcher project. In particular, they represented the ideological high points of her second and third electoral manifestos. The contours of this project were shaped by a wide range of measures within the field of urban regeneration: the centralisation of resource control and policy direction for local government, the privatisation of urban regeneration policy, achieved to some extent through the decentralisation of local government functions and the contracting out of local government services. At the same time the policy procedures for dealing with urban regeneration policy within the public domain were reconstructed and an attempt was made to reorientate public expectations of local government's role in local economic policy (Thornley 1991).

The apparent contrast between the urban regeneration and economic development of the 'old' (local Government) and of the 'new' (government quangos) was therefore stark. It appeared that where local authorities were bureaucratic, UDCs would dispense with red tape. Where local government was regulatory, the UDC would be entrepreneurial. Where the former was slow, and encumbered by participatory processes, the latter would be fast and effective. Finally, where local authorities were large hierarchical bodies, fraught with departmental rivalries and subject to a multitude of resource demands, the UDCs flat management structure and small staff would free it from unwieldy internal disputes.

There are now thirteen UDCs operating across the United Kingdom (1). Their impact and their success varies considerably (Imre and Thomas 1993). Although their organisational structure and strategy have a common root in legislation, these too have accommodated to the institutional and economic topography of their various locales (INLOGOV 1991).

Since their introduction the arena of local government has substantially altered, so transforming the institutional apparatus and the practices of urban regeneration policy. The rapid expansion of public service agencies has involved a reformulation of the actors participating in the delivery of regeneration policy and in addition, has altered the style of policy implementation. The fragmentation of local government functions has been accompanied by the centralisation of fiscal and ostensibly of political control (Gamble 1988). There has also been a move away from a locally representative accountability, towards a deregulated and privatised decision-making arena.

There is no doubt that local government has been transformed, superficially at least, to something resembling a local 'governance' (2). UDCs are just one element in this project of governance. They are subject to greater scrutiny by central government and more formidable intervention by private sector interests.

This institutional evolution within cities has paralleled a massive economic restructuring of Britain's urban economies (Massey and Allen 1988). These changes are the result of a transformation in global processes of production, consumption and exchange which have deeply penetrated locales (Storper and Scott 1992, Dunford and Kafkalas 1992). The restructuring has left urban areas poised between two 'states': dereliction and obsolescence arising from the departure of manufacturing industries designed on Taylorist and Fordist principles on the one hand; on the other, an embryonic future predicated upon new distribution functions, technological innovation within the professional commercial and services sector – high value added consumption activities and attractive locational attributes for command and control functions.

The evolution of the Development Corporation is therefore central to an analysis of the real meaning and impact of these events. UDCs provide a valuable insight into the broader processes which are recorded here. But the UDCs have themselves been the subject of intense scrutiny and have thrown up a variety of accounts of their significance to the contemporary world. Before moving on to consider my own interpretation, I would like to look briefly at some of these narratives and examine what they tell us about the initiative thus far.

Economy, Effectiveness and Efficiency?

Within the urban policy arena there has always been a concern with the effectiveness and the efficiency of the policy instruments that are used. This has become more marked in the present environment, as neo liberalism heightens public perturbation about 'waste' in government. Three major reports covering UDC activity have examined their performance using efficiency, economy and effectiveness criteria (Public Accounts Committee HC385 1989, NAO 1989 and NAO 1993). Their recommendations have shaped the later evolution of UDC policy. It is interesting to note in this context that because of the local conflict which the UDCs have generated and the lack of definition in the Act, both the Select Committee on Employment and the National Audit Office have been conspicuously interventionist and some would say Keynesian, in their recommendations for policy change.

In all cases, modifications to practice were the preferred solution in overcoming the inequitable effects of UDC policy. Furthermore the reports noted the lack of interagency collaboration and poor local institutional relations, but questions evaluating the institutional and processual approach adopted by the UDC within the urban arena were left unanswered. While therefore, these accounts tell us something of the way in which the UDC is measuring up to Treasury expectations, they go no further in explaining policy failures. Nor are they particularly concerned with the political implications of UDC activities, other than when politics is likely to impact on the administrative and managerial operations of the Corporations. Thus they perceive the policy in isolation from the framework that underpins it, and are not too helpful in answering such questions as: what impetus drives the policy? What interests prevent it from achieving its objectives? Or even, what impact is it having on wider city government? Overall, although the individual reports identify key areas of concern, none of them place the UDCs within the reach of the literature that has explored the problematic of policy implementation within the state. Nor indeed do they appear to treat the UDCs as a new type of urban policy instrument and thus to recognise the difficulties which the development of a particular type of new policy process may present them with. So we are left with a catalogue of policy difficulties, which are suspended in an organisational and political vacuum.

Social Equity versus the Market

Related to the work on the effectiveness of UDCs, are accounts which contrast the UDC with local authorities and which counterpoise the values of social equity or Keynesian values, against the market or neo liberal values, and correspondingly accountability and democratic decisionmaking against non accountable non democratic decisionmaking. These writers are preoccupied by traditional post war urban policy considerations, principally the negative effects of spatial or economic restructuring. They highlight employment, community benefit and the importance of local accountability and participatory structures as central concerns in the future policy development agenda of the Corporation.

The Centre for Local Economic Strategies (CLES 1989), points out that only in 1988, (7 years after the start of the initiative) did the Select Committee on Employment develop a definition of regeneration which included employment generation as an element within the UDC programme. This definition has been used subsequently by CLES to advance social equity considerations, but the Urban Development Corporations themselves have also used it to develop different strategies to those adopted by for instance LDDC. The 'bringing in' of employment to the Corporations' agenda can, in part, be seen as a victory for local campaigning groups, but it is also a vital readjustment of the policy in order to increase the legitimacy of the UDC and minimise conflict in the locality.

Robinson et al. exemplify the approach to community benefit within the social equity model:

> 'our particular concern is with how they relate to local communities and the twin issues of community participation and community benefit' (Robinson et al. 1993).

The theme is consistent with wider social policy concerns about the impoverishment of inner city communities and the approach which the state should adopt to deal with this problem. Connected to this is the suggestion that UDCs are reinforcing inequalities within regions (CLES 1989). Although inequalities within regions and within cities appear to be visibly more pronounced since their inception, the UDCs role in the further emiseration of urban communities (other than by commandeering urban programme finance, which in many cases would have been used in the same way) is difficult to verify. This is because there are many reasons for shifting populations and equally there are a number of possible causes for the intensification of inequalities of poverty and wealth. Perhaps the argument should be that the UDCs are iconographic of the institutional dimensions of social polarisation in the inner cities.

Aside from employment the overall thrust of this perspective is a concern with 'who gets what', in other words who benefits from the UDC? Who gets what however is a question that is caricatured through anecdotes about 'UDCs existing to benefit property developers'. Community benefit has come to be the bench mark by which the progressives have judged the UDCs activities.

That this should be a central issue in the narratives describing UDCs is not criticised here. It seems to me important in terms of fairness and equity that such questions are raised, and even more important that in terms of hegemonic politics, they remain on the agenda. Although these accounts point to the UDC role in reorientating the direction of urban policy away from equity preoccupations they do not develop the link between this and larger concerns about why they are pursuing this course of action.

A critical factor in this work is their juxtaposition with spatial concerns, specifically the identification of local sectoral peculiarities and the impact that these have on the type of model employed by UDCs. The CLES study cites the case of local land markets and explores how much less conflict there is in an area with no discernible land market, the obvious example being Liverpool contrasted with LDDC. Where there is a demand for land there tends to be conflict over use and exchange values and compensation for relocation. Like the general critique of the urban policies of the 1980s, Rebuilding the City (Healey et al. 1992), the Robinson report criticises the property focus of the UDC, primarily because 'property-led regeneration does not create jobs' and because it can 'distort the market'. The critiques of the property focus in UDCs are now numerous. This is hardly surprising given the downturn in the property market, the inefficiency of the trickle down concept and the inability of the policy to have any widespread long term benefit, but the question remains who benefits from property-led regeneration and why has urban policy moved in this direction.

Stoker had earlier pursued this argument, but he incorporated processual concerns into a social equity framework (Stoker 1989). He examined the social, economic and political outputs of the UDC in an attempt to ascertain whether the UDCs are really 'the most important attack ever made on urban decay'. Noting the differences between the generations of UDCs, especially in their relationships with local authorities he concludes his investigation by highlighting the weakness inherent in a property-led approach to regeneration, especially in relation to community benefit and employment. By looking at social outputs he observes the potential of a 'memorandum of agreement' in negotiating further gains from the UDC. When examining political outputs, he observes the political context growing up around the UDCs, in which they are cultivating a closer working relationship with other agencies in their areas, and the concern voiced by local politicians about who will get the 'credit' for what will take place and what it will do to political constituencies. Implicit in his work is a recognition of the importance of the democratic process vis a vis local networks; the expansion of networks, for Stoker, is problematic. His apprehension rests on the question of what the UDC will mean for the whole process of democracy:

> 'the other UDCs have tried to bed themselves within the complex regional network of civil servants, local business people, trade unionists and local government officials that predate their arrival' (Stoker 1989).

A key consideration for local politicians in evaluating UDCs is the impact that they have on the redistribution of patterns of urban finance. CLES argue that while UDC funding has increased so Urban Programme has declined. In addition, complementary finance (EZ, city grant, DOT monies) is not counted against the main UDC expenditure totals. CLES also suggest that leverage ratios as a method of measuring private sector investment are desperately uneven and particularly susceptible to local conditions and are therefore not ideal as measurable output.

These accounts are marked by the tendency to draw in equity considerations and contrast local authority practices in urban regeneration with those of the UDC. CLES, in particular, contrasts the local authority strategic planning and economic development approach with that of the UDCs, for instance criticising their reliance on master plans, area frameworks and public competitions, rather than local and structure plans and public participation techniques. The criticism extends to the marketing-led practices of the UDCs.

The difficulty with this critique is that although one can use an ideal type

UDC model and contrast it with an ideal type local authority, the actual practices of local authorities and UDCs are much less discrete.

While it is true that local authorities are compelled to be, and in many cases want to be, more accountable to local populations than UDCs, many have resisted participatory practices. Likewise local authorities are compelled to be more strategic, but often fail to be so because of territorial rivalries with neighbouring authorities (Healey et al. 1988).

Clearly the UDCs are less concerned with the equity considerations of regeneration policy than previous urban regeneration initiatives. Employment was grafted on to the aims and objectives of UDCs in the light of earlier critiques, community development was initiated as a response to the woeful attempts at local liaison by LDDC and as a more pragmatic vehicle of generating new alliances in localities. But it is nevertheless interesting to speculate why the UDCs should become enmeshed in precisely the same equity concerns as their predecessors. Many accounts of 'who gets what' in the regeneration benefits of UDCs rightly concentrate on the anecdotal and the local. The important question of equity in UDCs ultimately points us in the direction of the larger structural considerations which individual cases help illuminate. Commentators were interested in two issues here: the shift away from equity-led social and employment policy, and the movement towards properly-led activities. I argue here that the UDC was created precisely because it provided a more legitimate environment for this shift to take place. The UDC institutionalised this shift in British local government activity.

Central Local Relations

Inevitably the literature on the UDC is preoccupied by the subject of central local relations and a closely related field, the recent work on spatial differentiation, most pertinently the restructuring debate. Batley was one of the first to probe such issues by focussing the definition of the 'interests' represented in the UDC (Batley 1989). He suggests that the rhetoric and the practice of the Urban Development Corporation can be clearly distinguished from that of local government. He sees local government as powerful multi-purpose and accountable organisations and UDCs as uni-purpose and non-accountable, promoting central corporatism, operating locally. The dominance of some bodies and the support of others in local politics, depends he suggests, on the structures of institutional influence, in this case, their organisational abilities. Batley goes on to argue that UDCs are designed to evade planning regulations, and he suggests that local authorities are precluded from the role of entrepreneurial land development agency because of the importance of statutory planning legislation to their legitimacy. Batley also emphasises the increased control which government has over capital flows in the

UDC and the freedom which this gives Central Government to determine the nature of infrastructural investment. The UDC becomes a partner in this activity, rather than a regulating body. The powers of agglomeration are thus vested in the UDC depriving other public utilities of that same power. Nevertheless he argues that the UDCs are manifesting different developmental patterns because they are contingent on local relations and processes.

Batley presents a very powerful argument, in many ways reinforced by this study, but it is also true to say, as the Robinson study indicates, that the delineation between local government strategies and those of the UDCs is often unclear. While this idea is appealing, it is not entirely correct as local authorities have often been equally evasive where public regulations have hampered them in pursuing land development deals (Batley 1989). Furthermore his assumption that local authorities cannot be 'entrepreneurial' is questionable as many local governments developed entrepreneurial strategies prior to the establishment of the UDCs.

Spatiality and the UDC

The tension between the 'local' and the 'global' evident in the institutional literature on local central relations now encompasses the UDC narratives. The most recent work on UDCs has sought to integrate spatial logic into explanations of institutional form and variation. For example, the overarching concern of Imre and Thomas is how spatial variation is manifest in the practices of the UDCs:

> 'in particular it is increasingly clear that there is an imperfect understanding of the locational development and delivery of national urban policy, or how its wider objectives are sustained modified or contradicted by local socio-political milieu' (Imre and Thomas 1993).

The question implicit here, is – if there is spatial variation, why, when UDCs are so tightly integrated into Whitehall, and how, what are the processes that we are observing which tell us that it is space and geography which cause policy variation across localities?

Brownill answers the question by analysing the micro processes of social relations within the Dockland area over an eight year period (Brownill 1990). She attempts to find a method to link the global with the local. She concludes that 'an attempt at restructuring can make an area a hostage to the fortunes of the property market as well as working to the detriment and exclusion of the locality'. Ultimately Brownill's hypothesis suggests that temporal process assumes as much importance as the spatial factors in comprehending change. Spatial explanations alone are inadequate to the task of examining the UDCs. Spatial variation has affected the implementation of the UDC experiment, but what that effect is precisely and how spatial connects to processual, particularly state factors in policy development is as yet not fully explored by the literature.

Meegan comes closest to providing a satisfactory spatio-temporal explanation of the UDCs via the political economy of David Harvey. He draws on Harvey's conceptualization of 'entrepreneurial strategies' which are promoted by urban class alliances in order to deal with the vicissitudes of crisis-prone capitalism. Meegan examines the strategies of the Merseyside Development Corporation. He concludes that they do indeed draw upon the range of entrepreneurial strategies which Harvey depicts, but the origin of the strategies does not fit Harvey's theory, as Meegan cannot find any recent evidence of an urban class alliance. In fact the divisions between outer and inner Liverpool and the lack of a 'centred' local authority foundation for alliance activity stresses the fragility of MDCs approach (Meegan 1993). Meegan comes closest to my own account though I differ from him in the way that I use Harvey's formulation of structured coherence. While Meegan draws upon entrepreneurial strategies to assess the level of coherence or otherwise in the strategic configuration of UDCs, I see the strategies as an output of the urban class alliance, they do not determine the coherence of the urban region. Rather this is constituted through the triangle of technological mix, organisational form and social relations and it is through an analysis of the interpenetration of these factors that I move towards a conceptualization of the relationship between coherence, the urban class alliance and the UDC.

Restructuring

The links between economic and state restructuring are tentatively suggested by a number of commentaries on UDCs. Although some of these draw on a post Keynesian framework (Meegan 1993), and many use a general concept of Fordist and post Fordist practices (Stoker 1989), it would be difficult to describe them either as accounts of post Fordism, or as flexible specialisation. This is because accounts tend to draw upon both of these classifications as 'influences' rather than as examples featuring the necessary characteristics of each of the typologies. Within many of the accounts the distinctiveness of each approach merges into a general categorisation of a transition in regional economies and the institutions of the state.

There are two approaches to the accounts of UDCs within the context of restructuring. There are those who propose that UDCs are part of the mechanism which deliberately participates in the restructuring of localities, primarily economic restructuring (Batley 1989). There are also those who imply that the UDC has been introduced as a result of the restructuring that has taken place in the state form (Adcock 1987). To some extent my own conclusions bear out both of these positions but it is difficult within these accounts to reconcile the definite parallels between restructuring in the state and restructuring in local economies without having an overall view on exactly how the two positions meet.

The economic reorientation of the locality emanating from UDCs has become a particular feature of interagency conflict within the locale. For instance, Imrie and Thomas (Imre and Thomas 1993) illustrate how small industry suffered as a result of UDC policy and actually provided a focal point of opposition to the UDC. They point out that small firms affected by the plans of the Corporation were so alienated by the process that they eventually established their own small business forum. Bristol Development Corporation's experience over the same issue, the relocation of industrial uses in the centre of the UDA and the promotion of marketing and high tech uses in their place, is documented by Oatley (Oatley 1993).

Similarly, Byrne suggests that TWDC's project has been to wipe out maritime manufacturing for the Rivers Tyne and Wear. (Byrne 1989, 1993). Byrne perceives the UDC as a post industrial Trojan Horse explicitly rejecting a maritime industrial future within a global world opting instead for a property led strategy, consistent with strategic capital interests. For him the UDC is a 'colonial administration' which rejects the industrial past and predicates its future on ephemeral commercial interests.

Sadler places the new agencies of local governance, including the Urban Development Corporations, in the context of a transition to a new style of Corporate management:

> 'A key distinction of Britain in the 1980s was a redrawing of the legitimate extent and nature of state intervention in the workings of the market economy. Deregulation, intended to enhance the power of capital depended centrally not just upon creating new labour market conditions but also developing an environment in which capital has the freedom to invest and manoeuvre. One mechanism for the latter was a redefinition of planning epitomised in the creation of the UDCs'(Sadler 1990).

He illustrates his thesis with reference to the post war industrial restructuring that took place on Teesside where agreements were reached between the local authority and local firms. The new arrangements have changed completely and permanently the social foundation upon which planning rests and suggests that the new agencies prefigure the extension of corporate control over the political institutions. While the 'present' i.e. Fordist/Keynesian, formula represents one faction of capital state domination, what we see happening now, represents another. If this is so we should be seeing the same thing happening in all localities, do the new structures move around and if so do they change?

Here finally the Regulation literature represents the latest development in elucidating institutional change associated with the reorganisation of the state (Peck and Tickell 1994). Regulation theory is preoccupied by capitalism's ability to continually reproduce itself in the face of numerous internal contradictions and crisis (Aglietta 1979, Lipietz 1987 and Overbeek 1990) Regulationists conceptualise economic, political and ideological relations, by using the concepts of a 'regime of accumulation', within the economic sphere, coupled with a corresponding institutional and societal 'mode of regulation'. The regime of regulation is explained as a combination of production and consumption activities which fuel the accumulation process. The co-evolving mode of regulation is contoured by the problematic of class struggle, a struggle which is defined by the relations of production, circulation and exchange. In some senses the regulationists work in from the macro to reach agencies like the UDCs, rather than focus on a specific agency of regulation.

Therefore what I want to do in this book is bring together the various critiques, with my own empirical and theoretical investigation in order to conceptualise the overall pattern of change and interpret the elements of continuity, in other words to explain the overall uneven development of policy processes within the state form. The processes of policymaking in the UDC offer a unique window with which to conceptualise change and the motivation propelling that change against the backdrop of local global relations. The question remains however how are we to conceptualise and theorise changing state policy processes and their conceptions to broader political and economic topography of urban locales? This rich field of UDC literature, thus far has been unable to provide us with a theoretical review of the emerging processes of policymaking therein and no significant link has been made between the processes of policymaking and state change. The literature has been concerned to a significant extent with outputs and this emphasis has consigned much of the analysis of the UDC to conceptualising the project within the confines of the 'Thatcher project'. Of course any comprehensive analysis cannot be without such a component, but in some sense these interests are merely a facade on the architecture of state change. The UDCs fundamental importance lies not simply in the redistribution of the resources of public policy, the fundamental importance I argue is found in the embryonic change in the state form which they presage and the struggle that is taking place over that form.

The Conceptual Framework

> 'It is noticeable for example that in those European Countries in which municipal socialism has already won its laurels and where a more articulate

> class based politics does exist that the corporatist powers of the urban based class alliance are whittled away and replaced by the powers of the nation state where the bourgeoisie can more easily retain power' (Harvey 1985a).

My book attempts to remedy some of the existing gaps in the explanations for the UDC by developing David Harvey's theorisation of the existence of a 'structured coherence' and concomitant 'ruling class alliance' within the urban region. I will suggest that the development of UDC policy can be related to the level of 'structured coherence' within the urban regions and that a deeper analysis of this policy-making environment can reveal the extent of interdependence between the local economy and the local political milieu.

The tendency to structured coherence, is the tendency towards an equilibrium which sustains capital accumulation in spite of huge shifts in the technological mix, social relations and organisational form within an urban region. In other words, it is capital's ability to sustain continuity within the social relations of production in the face of enforced or necessary change. The role of the class alliance is to ensure that coherence is protected and enhanced. I want to explore the plausibility of this theory by applying it to an analysis of the Urban Development Corporation.

I want to extend Harvey's concept to incorporate an analysis of state relationships and the role that they perform in mediating institutional change. Harvey himself, makes only passing reference to this, and other writers who have used the concept, most notably Painter who has concentrated on local authorities and Meegan, have not had the opportunity to make systematic regional comparisons of these new institutions of local governance (Painter 1991, Meegan 1993).

Harvey argues that a 'spatial fix' is central to the process of capital accumulation within uneven development. I want to incorporate the methods of policy analysis research into the study of structured coherence, in order to emphasize the importance of the 'process fix' within contemporary state and capital relations. The 'process fix' I suggest, is the ability to change and adapt the process of policy-making within the state in order to complement changes which might become necessary in the regime of accumulation.

Applying the Concept

How could we conceptualise the UDC within such a framework? One could use the twin concepts to speculate that the 'structured coherence' which underpins the 'class alliances' in urban localities is under attack from the nation state. This process would involve the breaking up and reformulation of the regional networks involved in urban government. To a certain extent this can already be detected in the struggles which have taken place over local government's economic capacity during the period of the late 1980s (Boddy and Fudge 1984). The rationale for such an onslaught, can be found by conceiving of the nation state and the locality as representing different formulations of class fractions.

It could be argued that urban class alliances have been popularly, though unevenly, captured in the mid 1980s by oppositional interests in many local governments. The key to local government's power, lies, as Harding has suggested, in its role as a rentier and the transparency which public sector decision-making occasions (Harding 1992). Aside from this, within such circumstances, the demeanour of the class alliance in the urban region is crucial to the impact of the reorganisation of the global economy in the locale.

During the 1980s, the level of local 'structured coherence' showed considerable diversity. Many regions were weighed down by Keynesian processes and labour relations which implicitly held back innovation to fuel new accumulation strategies. In some regions, vested interests created a monopoly situation which led to economic stasis. Here the state sought to intervene in order to destabilise such interests, and produce demonstration projects, prompting a change in the state form and the coherence. Other regions were already involved in struggles as a result of the massive reformulation in social organisation that had taken place alongside economic and structural transformation. In other regions the state has acted to protect or advance the interests of certain capital fractions against the hegemonic momentum of opposing class forces, hence the development of many opposing models of economic and land-use planning which have emerged throughout the decade.

The UDC can be seen as part of a hegemonic apparatus for achieving the reformulation of the state in locales, and thus the level of economic coherence therein. This reformulation can be seen as symptomatic of the changing 'mode of regulation' in the evolving state form. The way in which relations with Urban Development Corporations have subsequently unfolded in the urban regions may reflect three factors:

A. The nature or characteristics of coherence locally.
B. The position of the urban class alliance.
C. The extent to which struggle and changes created by the break up of Keynesian corporatism, were taking place already.

The Development Corporations limit 'autonomous politics'. Autonomous politics are synonymous with the Keynesian Welfare state. Therefore the struggle between the state and the urban regions over Corporation policy and practice can be seen as a struggle about the dominant 'mode of regulation' which will emerge in the current period. In other words the UDC is not part of a central local dispute as much as a dispute over hegemony and regulation. Locality is crucial to our observation because it shows the spatial impact on public policy and correspondingly adds a temporal dimension by allowing an analysis of historical social formations. I suggest that the struggle which we are observing is about the form as well as the content of state decisionmaking. Attempts are being made to frame a way in which actors gain access to the decision-making arena and thus to the regional 'class alliance'. Effectively UDCs are a way of removing opposition in the locality and delegitimating local actions. In this context the policy process becomes the mode of engagement between those arguing for transformation and those resisting it. It is the nature of 'structured coherence' and the position of the urban class alliance which governs the response of the locality to the Urban Development Corporation and that by observing the UDC one can detect the linkages between changing micro processes at the local level and global economic restructuring. It is also suggested that the social relations involved in this restructuring are woven through the construction of the practices and procedures of urban governance in the locality.

The task of this book is to explore the connections between the strategies and processes which have emerged from the Urban Development Corporation and the impact that the locale has had upon its development. The objective is to make some sense of the transformation that is taking place in the modern state and the nature of the relationship between institutional and economic change in the locality.

Structure of the Book

Chapter one will examine the legislative and organisational framework of the UDCs. The legislative and political framework will be illustrated and the phases of Development Corporation activity discussed. This is contextualised with reference to the political and policy changes which have emerged concurrent with and from within the UDCs. In Chapter two I will examine the concepts of class, state and theories of state and policy change before finally moving on to explore how these concepts can be theoretically subsumed within David Harvey's notion of structured coherence and the ruling class alliance. In Chapter three I develop a conceptual framework in which I place the development of the UDCs. Chapters four, five and six document the detailed history of the UDC and its local policy development and political trajectories in three urban regions located across the UK; the Corporations in Bristol, Trafford Park and Tyne and Wear are analysed over a four year period. Chapter seven allows me to provide some concluding comments on the results of the empirical work.

Notes

1. This does not include the Belfast Corporation.
2. This term is now popularly used to describe the range of governmental, quasi governmental and partnership bodies associated with the governing of cities.
3. This should be remedied with the publication of the results of the ESRC local governance initiative findings which are about to be published. Within this initiative there has been substantial discussion about the applicability of Regulation Theory to the analysis of the state.

1 Overview of UDC policy

Introduction

> 'How much stronger the whole of society would be if the municipal endowment of the 19th century, which was essentially about capitalism and local government working together – because, of course it was run by the same people – was continued into this part of the 20th century (The Development Corporations are not only) an engine of industrial change...they are also a means of securing a social and more especially a political transformation'
>
> (Michael Heseltine, *Municipal Journal* October 1986).

The UDCs are now near to the end of their indicative lives. They have been the subject of three major government reviews (HMSO 1989, NAO 1989 and NAO 1993) and a substantive book charting the variety of institutional forms which embody the nationwide project (Imre and Thomas 1983) and a host of reports, compiled by voluntary, statutory and campaigning agencies (for example CLES 1991). All of the accounts of the aims and the activities of the UDCs reflect the incoherence of the political and indeed the physical and economic project, in which they are engaged. While however, there is little disagreement about the overall relative lack of UDC success (though there are some exceptions), even after the introduction of performance indicators in the early 1990's, there is no real agreement as to what constitutes a successful UDC. There is widespread variation in the policies implemented by the UDC and there is significant variation in the extent of their embeddedness in the local political networks. So what are the UDCs? How can we judge them?' Are we simply to read them as a new vehicle of regeneration? Or do they represent an extension of the ideological project of privatisation and agentification? (Saward 1995). Finally what do they mean in terms of the changing institutional form of the city?

Urban Development Corporations do indeed represent a shift in the state's approach to urban regeneration. While they have some features in common with previous agencies of urban regeneration, their particular combination of powers, functions and resources, married with a New Right ideology, has altered the template for the regeneration policies which have hitherto been typically employed in cities, by British local government (Potter 1988). But an understanding of their overall national political significance must be tempered in the knowledge that the impact which spatial factors and their historical precedents have upon the state form, is profound. The Development Corporations are not a homogeneous policy entity. Though they might have enormous institutional similarity (Imre and Thomas 1993), the differences which are emerging between them can be explained by the argument that they are at the centre of an ideological and political struggle within the context of dramatic social and economic change in the UK. There are three factors that are particularly significant in terms of the struggle which we observe developing around the UDC. Firstly the gradual dominance of the Thatcherite element in the Conservative party during the 1980's, whose specific political agenda was significantly coloured by the variation in the economic cycle underpinned by longer term structural economic trends. Secondly the impact of locality based struggles over the implementation of Corporation policies and finally the impact of oppositional strategies around local economic development during the 1980's and early 1990's.

The chapter is divided into four sections. The first section contains a description of the legislative and policy framework for the UDCs. Section two describes the implementation of the UDCs, the characteristics of each Corporation, the political responses to their imposition, the relations which developed around them and the national political context in which the Corporations have evolved. Section three will look at the opposition to UDC policy. The concluding section of this chapter will evaluate the overall impact of the UDC in the context of contemporary urban policy.

The Framework

The thirteen Urban Development Corporations are single issue agencies charged with regenerating their areas within a limited time period. They employ small permanent staffs and are run by a part time board of Directors, appointed by the Secretary of State, but recruited from the locality. They are responsible directly to the Secretary of State for the Environment but interact with a number of regional, governmental and other statutory and non-statutory agencies responsible for the regeneration of their area. Neither the UDC, nor its powers, aside from vesting, are entirely new concepts in British urban policy. Development Corporations have a long line of antecedents on both sides of the Atlantic (Barnekov, Boyle and Rich 1989). But the amalgamation of powers under the direct control of central government is novel, as is the way in which the UDC has gradually developed as a political weapon in the public policy arena. Urban Development Corporations signal a significant change in the direction of traditional Keynesian state approaches to urban policy (even though there have been substantive changes in policy direction within this field during the post war era).

Firstly UDCs are market led rather than needs led, overturning traditional equity notions of urban policy. They act upon supply (land) asymmetry, whereas Keynesian state policy argued for stimulating demand (Healey et al. 1992). Secondly they have moved away from policies focused on labour market strategies shifting their emphasis to land and capital, indicating a move towards property-led regeneration (Healey et al 1982). Thirdly, they are explicitly committed to the promotion of private sector leverage rather than public sector subsidy (Brindley, Rydin and Stoker 1989), and their substantial preoccupation in many localities involves shifting local populations rather than redistributing transfer payments as a means of targeting 'problem areas' (DCC 1990). Fourthly they have economic rather than social policy imperatives (Imre and Thomas 1993). Fifthly they have popularly promoted consumption rather than production related strategies (though this is subject to particular uneven development) (Imre and Thomas 1993). Finally their ethos is anti-bureaucratic and initially at least they were not in favour of public participation in their policy-making procedures (1). All of these factors mark out the UDC as an important institutional, economic and political experiment, differentiating them not only from the preceding urban initiatives, but also from the initiatives which have developed subsequently.

The Act

The local Government Planning and Land Act of 1980 was the legislative foundation for the subsequent establishment of all the Urban Development Corporations. The Act contained 19 parts and 34 schedules. It had a problematic, though speedy gestation. Firstly in the cabinet; Michael Heseltine has described both the Treasury's opposition to the UDCs and his own arguments with his cabinet colleagues, who retained a commitment to Conservative election manifesto pledges to reduce Quangos. Secondly it had a difficult passage through Parliament, partly because of its sheer volume and partly because it was contentious, as it initiated the series of changes which were enacted throughout the 1980s to control and limit local governments' independence in local financial control (Heseltine 1990). The Act presaged a sophisticated reorganisation of the institutions, powers and practices of the contemporary state (Gamble 1988).

Powers

The objectives of the Urban Development Corporation as set out in the Act were to:

- Secure the regeneration of its area by bringing land and buildings into use.
- Encourage the development of existing and new commerce and industry.
- Create an attractive and social environment.
- Ensure that social and housing facilities are available (Moore and Catchpole 1983).

Under part xvi of the Act, the Secretary of State for the Environment could establish an Urban Development Area 'where it is expedient to do so' (Moore and Catchpole 1983). The implementation of the Act's objectives was based on a combination of a series of local authority powers and functions that would be transferred to the new agency: those of land use and development, development control and industrial development. Within local authorities, these functions were generally administered under three, often separate, departments – the Planning Department, the Estates and Property Department, and latterly embryonic Economic Development Departments. Amalgamating the powers in a site specific agency was intended as a policy solution to the interminable interdepartmental negotiation over economic development projects which it was alleged took place in local authorities.

Planning Powers

Under the Act, the Urban Development Corporation becomes the Development Control body for the Urban Development Area, thus divesting local authorities of their powers in those areas. This means that it can:

- Determine planning applications, grant deemed planning consent to carry out development anywhere in the designated area.
- Deal with planning blight under the normal provisions of the Town and

Country Planning Act 1971, and the provisions of the Land Compensation Act 1971 also apply.

- Deal with conservation areas, orders and preservation issues within the UDA.

There are formal consultation procedures, (usually limited to fourteen days), for dealing with local authorities and other relevant agencies about processing of planning applications and development proposals. The Development Corporation is subject to the statutory planning documents for the area, that is, the District, Structure and latterly the Unitary Development Plan. There is, however, a provision in the Act to allow the Secretary of State to authorise a Special Development Order to be put before Parliament if necessary. This permits the Development Corporation to become the statutory planning organisation within the UDA, and thus acquire plan making powers. Apart from the incorporation of the New Towns, the divestiture of planning authority from local government for areas of land within their domains, again signalled a departure from the ethos of planning laid down in the 1947 Town and Country Planning Act.

Land Powers

The UDC can "acquire, hold, manage, reclaim and dispose of land, within its area". The same provision applies to property. Additionally it can:

- acquire land adjacent to the UDA
- acquire land adjacent or not which it requires to discharge its functions
- dispose of land in any way it thinks fit for a sum lower than the costs of acquisition.

UDCs can voluntarily acquire land but also have the option of compulsory purchase and in section 141 of the Act they have the power to request the Secretary of State to place Vesting Orders. Vesting transfers local authority or public land and its freehold to central government, and is considered to be critical in land assembly. This is perhaps the most significant element of the panoply of powers conferred on the UDC and arose from central government's enduring concern with the expanding local authority landbanks, accruing from the demise of industrial infrastructure in the urban areas. It was significant too in that it allowed UDCs (in effect central government) to 'pick up' land in inner city areas and develop it according to criteria outside the normal planning guidance for the area, thus giving central government considerable control over previously local government held land.

The final set of interlocking powers which are held by the Development Corporation relate to industrial development. These powers are derived from the 1978 Inner Areas legislation and allow the Corporations to:

- provide loans for acquiring land
- provide loans and grants for co-operative enterprises
- provide loans and grants for improvement areas
- provide loans or grants for building on any land sold or let by the Corporation.

Organisational Form

The UDC is one of a growing number of arm's length agencies which were established for the purpose of undertaking previously municipal functions by successive Conservative governments during the 1980s (Weir and Hall 1994). They have been termed quasi governmental arm's length agencies or 'non elected local government' (Stoker1988), because previously local authorities typically implemented all or most economic development and training functions implicit in urban regeneration activity, in house. These functions have now been transferred to a number of external agencies, such as Training and Enterprise Agencies, Development Corporations or urban Partnerships. Local authorities previously established joint boards or interagency management frameworks which reported back to sponsor bodies, for the implementation of economic policies. Not so the new agencies. Characteristically these are free standing, remaining separate from both the regional Departments of central government and from local authorities, though often, local councillors sit on their boards and their officers participate in joint policy fora. In the case of Urban Development Corporations, councillors are nominated in a personal capacity, however many other urban regeneration partnerships allow official nominations from local Councils (Harding 1992), this is a feature of the later City Challenge initiative (DOE Guidance Notes1992).

The DOE designates the UDC as 'sponsored organisations' or 'non departmental public sector bodies' (DOE 1993). Ostensibly at arm's length from the state, they are in fact tightly controlled by Whitehall. Regional departments of central government oversee any significant financial transactions on behalf of the UDCs. This again gives Whitehall greater powers of intervention over particular economic policies and regeneration initiatives in far flung urban regions. The Secretary of State is responsible for the appointment of the Chairman of Development Corporations. In consultation with the Chairman, the other members of the board are also appointed by the Secretary of State, the Board number's between five and eleven persons (Moore and Catchpole 1983). The criteria for their appointment remain somewhat ill-defined:

> 'The Secretary of State shall have regard to the desirability of securing the services of people who have special knowledge of the locality in which the UDC is or will be situated and the Secretary of State shall consult such local

authorities as appear to him to be concerned with the regeneration of the urban development area' (Moore and Catchpole 1983).

The boards are dominated by the private sector, comprised of primarily commercial and professional representatives and in some regions industrial representatives. The public sector is usually represented by local councillors, ex MPs, trade unionists or academics. The proportions of each sector show regional variations with some important material consequences as I shall demonstrate in later chapters. Typically the boards meet every six to eight weeks, though some Corporations now have a monthly cycle. The members are part-time and paid a salary for their attendance. The selection of the board, as opposed to the election of local councillors, means that the lines of accountability for decisions taken by the UDC differ substantially to those of an economic development committee within a local authority. And the fact that constitutionally, they meet in private with no one other than board members or UDC personnel, and are not subject to the Local Government Information Act, means that the capacity of local interests to intervene in the regeneration process is highly selective.

The Corporations are small, retaining a minimal staff. LDDC has the maximum staff at 269 (DOE 1993). A 'typical' structure for the organisation appears to include four departments, a Finance and Administration directorate, a Development directorate, a Marketing, Public Relations and Community Development directorate, and a Commercial directorate. The level of staffing has had two consequences for the Corporations – the extensive use of planning and management consultants in the development of policy proposals, and the necessity of developing agency agreements with local authorities in order to process planning applications.

Funding

As sponsored or non departmental public bodies the Corporations are subject to financial scrutiny of their sponsoring department, the DOE. They are allocated financial support through the mechanism of the Annual Public Expenditure Survey. Each Corporation has an annual External Financing Limit comprised of three sources. They receive grant in aid for non commercial expenditure and a small degree of borrowing from the National Loans Fund. This is the Development Corporation's borrowing requirement. Both of these funds are controlled centrally and have in recent years been used to discipline the Corporations. In addition they also now attract European Commission finance (DOE 1993). The NLF requires that revenue flows from any scheme should be sufficient to meet the remaining costs. A real rate of return of 5% is expected on the commercial element of the UDC programme. Each year the UDC is required to submit a corporate plan containing a five year expenditure summary, which the Secretary of State must approve before the finance is released. These include performance indicators, which have recently provoked some dissatisfaction (NAO 1993). In addition to this, there are two other annual meetings between Chief Executives and Whitehall in every year as well as a degree of negotiation with Whitehall over individual projects (2). Many Development Corporations have found it necessary to re-negotiate the sums of money which are routinely and exceptionally released by regional offices of the DOE (Interviews with executive officers).

Figure 1.1 Creation of the UDCs

		Date Created
First Generation	London Docklands	1981
	Merseyside	1981
Second Generation	Trafford Park	1987
	Black Country	1987
	Teesside	1987
	Tyne and Wear	1987
Third Generation	Central Manchester	1988
	Leeds	1988
	Sheffield	1988
	Bristol	1989
Fourth Generation	Birmingham Heartlands	1992
	Plymouth	1992

Source: DOE Annual Report 1993

Politics and Place: The Introduction of the UDC into the Urban Regions

Phase One 1981-1986

Michael Heseltine explains the genesis of the first Corporations thus:

> 'what we perceived in 1979 was that because vast public expenditure programmes had been running parallel with urban decline, simply making those programmes a little more generous did not address the fundamental

issue of decline. Indeed it might have even accelerated it by perpetuating old attitudes and old policies... we decided to reverse the thinking... the basic idea was to invest rather than fund... the theory that this interventionism would make inner city areas competitive' (Michael Heseltine in the *Financial Times* 30/10/90).

The first generation of UDCs was established in London and Liverpool. They were given a fifteen year life span and access to unlimited funds, subject to public expenditure limits. They operated in entirely different contexts – LDDC next to the City, one of the biggest financial centres in the world, and Merseyside Development Corporation, within the depressed region of Merseyside. LDCC covered five London boroughs and dealt with another sub regional authority, the GLC, while Liverpool dealt with three Merseyside districts and Merseyside County Council (both county councils were abolished in 1986). London occupied a vast 2070 hectares. It had a significant population – 40,400 – and a working population of 27,213 (Docklands Consultative Committee 1990). It was a housing authority and contained an Enterprise Zone. Furthermore the process of planning and collaboration over renewal was not new in the area. It had been the subject of local and national government interest since 1972 (Burton 1986). By contrast Liverpool covered a much smaller 350 hectares (though it was expanded to 865 in 1988). Liverpool was more fragmented, politically and strategically, it was separate from the rest of the city and there had been no prior detailed redevelopment plans for it (Meegan 1992).

The stories of London and Liverpool are well documented elsewhere (Adcock 1986, Brownill 1990), but here I will outline their key emphasis. Their boundaries contained 'substantial amounts of derelict and under used land' and their aim was to encourage 'maximum private investment'. Their focus was on physical renewal, leveraging private sector finance, developing 'non planning' styles of public sector intervention and promoting entrepreneurial ways of working (NAO 1993 HMSO 898). Yet they were strikingly different in implementing their policies. It took LDDC until 1987 to overcome outright conflict with community groups and local authorities in the docklands area (Brownill 1990). Their secrecy and inability to consult on major developments became a significant local complaint. Merseyside on the other hand developed collaborative ways of working, and while they were more successful with the local community than with Liverpool City Council between 1983 and 1986, relations have since been relatively harmonious with all the local authorities in the area.

Phase Two 1986-1987

The next phase of Urban Development Corporation activity was announced at the Conservative party conference on 8th October 1986. Nicholas Ridley ushered them in saying:

> 'We won't let Labour extremists stop us. We know what they do with local power and local power is a stepping stone to national power'.
> 'Decades of socialist council rule combined with changing patterns of industry have left them (the inner cities) in a wretched state' (*Financial Times* 9/10/86).

This rhetoric gives some impression of the scale of industrial dereliction experienced as a result of economic transformation in the inner cities. But it also suggests, the political impact which the alternative strategies for renewal promoted by Labour local authorities were having on the Conservative Party.

Four Corporations were announced in this second phase: Trafford Park comprised 1,215 hectares of land around derelict docks and canal in greater Manchester. It was situated in Trafford Park Industrial Estate facing the ambitious Salford Quays development established in 1981 by Salford City Council. Trafford had almost no population at the time of its designation but significant employment at 24,468. Trafford Metropolitan Borough Council argued for their 'own particular type of Development Corporation' with expenditure of £90 million over a five to ten year period. They already had an Enterprise Zone within the proposed designated area and many of the modern industries located in the Industrial Park were integral to the multinational development strategies of Fordist companies (Trafford Park Metropolitan Borough Council 1990).

Teesside was a huge 4,858 hectare site of derelict land on either side of the River Tees and estuary, with an offshoot at Hartlepool. The departure of the British Steel plant at Redcar, the simultaneous decline and restructuring in the chemical industry and the death of northern shipbuilding had left large amounts of toxic waste land on the banks of the Tees. The Teesside UDA does not have any population and neither does it have an employment base within its boundary (INLOGOV 1989).

The Black Country Development Corporation is another substantial site – 2598 hectares – housing a significant population of just over 35,405. The Black Country boroughs had previously published a three year programme of derelict land reclamation (Nevin 1991). The Black Country has the largest amount of derelict and contaminated land of all of the entire generation of Development Corporations. It also contained the highest number of jobs: 53,000.

Finally, Tyne and Wear is focused upon a narrow strip of land on either side of the banks of the Rivers Tyne and Wear, encompassing 2,375 hectares. It houses a population of 4,500 and has an employment level of 40,115. It has been suggested that TWDC would provide a strategic overview of the derelict and active industrial and shipbuilding sites in the area following the abolition of Tyne and Wear County Council (*Financial Times* 9/10/86).

A further Corporation was established at Cardiff Bay and this was announced a little later in April 1987. This site surrounded the docks area, separate from the city centre. It was slightly smaller than the other Corporations at 1,093 hectares, and its population was around 500 with a working population of 1,500 (Imre and Thomas 1993).

The social economic and political context for each Corporation was different, their size and the scope of their activities varied, they contained different populations, varying levels of housing and employment, and developed a range of regeneration strategies. They dealt with differing numbers of host local authorities with diverse politics, but their powers were similar. They were given a limited life span – 7-10 years – and a budget of £160 to £200 million to spend (*Times* 24/11/86). They were all placed in areas of manufacturing decline which contained substantial derelict land within or near their boundaries. These sites, of 2,000 hectares or more, were to be used to develop new industry and commerce. The economic arguments for the establishment of this phase of corporations was very strong given the maelstrom of economic dereliction contained in the constituent regions.

Phase Three 1987-1994

The third phase of Development Corporations was significantly dis-similar to the other two. Three 'mini' and 'cashless' Corporations were announced in Leeds (375 hectares with a population of 800), Manchester (187 hectares, a population of 500 but a huge working population of 15,300 in proportion to its size), and Bristol (420 hectares, again with a large employment level of 19,500 but with no resident population) (*Financial Times* 24/11/86). Their life span was even shorter than the previous generation, between 5 and 7 years, and they each had £15 million pounds to spend. They shared four characteristics which marked them out from the other Corporations, they were much smaller in area and were established in areas with reasonable economic prospects. They had to be single-minded in their approach to unlocking the development potential of their areas and they were definitely short life agencies. They were all located close to commercial or financial city centres. In fact they were all placed in areas that promised significant growth potential. This clarified the implicit intention of earlier Corporations to move urban policy away from a concern with widespread deprivation towards the consolidation of narrow economic interests (Robson 1988). Together with announcement of the mini 'cashless' Corporations, there were two other announcements. The first was the intention of setting up a 'Midi'-Corporation in Sheffield. This was to be 900 hectares and was to last for seven years and was given a budget of £50 million. At the same time Merseyside and Black Country had their boundaries extended.

Phase Four

Two Corporations have been established since the third phase – one in Birmingham Heartlands, which earlier had been the subject of special development agency status, negotiated in a special deal with Secretary of State Ridley. At the time of the announcement of the second generation Heartlands was to be 'left to the Labour controlled city council' (DOE 1991). Plymouth was also announced and was established on the demise of the naval dockyard in the City. It was rumoured that the money made available for the UDC would equal that of naval disinvestment in the area.

The UDCs then appeared to represent a significant departure from previous urban regeneration initiatives. Their organisational structure appeared to preclude public representation and participation in local economic development and planning policy. Rather they were tied more closely to Whitehall through their financial regimes and to selected members of the public sector and private actors in the locality through a range of informal processes. They were encouraged to be entrepreneurial and creative, operating according to very broadly defined criteria: how their success would be measured was still very much a contested area.

Local Central Relations and the Context for Implementation

When asked about the motivation behind the Dockland experiment, Michael Heseltine stated:

> 'These people (the councils) have created an environment which is wholly unattractive to a balanced society, but then there is no question of them ever controlling it. The politics of the areas will change. Once you have a balanced community they won't be able to get away with this nonsense. It's a political and social revolution. Part of this revolution is achieved by decanting families of council tenants out of the urban estates to make way for new development and young economically active families (*Municipal Journal* 24/10/86).

The first Corporation was introduced against a background of significant public sector expenditure cuts, economic recession and, in Liverpool, inner city riots.

Indeed the Secretary of State for the Environment himself opposed public expenditure cuts of £5000 million pounds at a cabinet meeting in July 1981 because of his inner city brief (Young 1989). Not long after this he visited Liverpool's Toxteth, taking with him a charabanc of private sector bosses in the hope of igniting their sense of social responsibility. This resulted in the development of the first of many public/private sector partnerships that were established during the 1980s and effectively set the tone for the urban regeneration policy of the Thatcher years (Heseltine 1990). The Financial Institutions Group and later British Urban Developments, subsequently played a prominent role in promoting a new ideology of regeneration and an innovative approach to the task of developing the inner cities. Despite the malaise induced by the recession of the early eighties this group and their collaborating civil servants clearly felt a surge of optimism about the new way of doing public regeneration, even if at the time they scarcely had a hint of what it would mean for them or the work it would entail for their companies (Healey et al. 1992).

The political environment into which the second generation of UDCs arrived, differed enormously. Electoral rhetoric was just one factor which prompted Nicholas Ridley to suggest that he saw an even stronger future for such agencies. Alongside the new Development Corporations, the late Lord Ridley signalled a major change in local government finance, as a prelude to the introduction of the poll tax (*Financial Times* 9/10/86). This announcement arrived hot on the heels of a series of major clashes between local and central government over the powers of the former, and at the same time as the abolition of the GLC and the Metropolitan County Councils. Moreover the mid eighties had witnessed a series of public confrontations between local and central agencies about the scope and nature of state sponsored economic regeneration activities (Wainright 1989). The Conservative government was extremely hostile to the ideological manifesto of the leading urban 'left' authorities and their approach to economic restructuring. The approach echoed by many local authorities, entailed creating and promoting a series of innovative alternative plans and agendas for economic development (Boddy and Fudge 1984). Two issues dominated this discourse. The first concerned the substantive response to economic restructuring – many local authority plans sought industrial, manufacturing and production based development aimed at broadening the skills base and enhancing the job prospects of the local population through training initiatives. The government however, promoted consumption and service based strategies, hinging on the clearance of the debris of manufacturing industry. The second issue was how the state should approach reorganisation of the economy, who should be involved and how participation should be structured. Local governments opted for widespread common ownership of the vehicles of regeneration, either through nationalisation or local common ownership and workers' control structures.

Central government wanted to limit local government's involvement and participation in the implementation of strategies and devolve ownership to various elements of local capital.

The flavour of the opposition to the second generation of UDCs also differed from that of the first phase. Opposition from national political parties emphasised accountability and representation. The Liberal Alliance election manifesto announced its intention to democratise UDC boards and integrate them into regional representative structures (*Financial Times* 13/7/89). Labour's then Trade and Industry spokesman, the late John Smith, described the Corporations as 'undemocratic and centralist'. He linked the UDCs to other aspects of the Conservative planning and industrial agenda, reiterating a commonly held fear of the British trade union movement about the orchestrated decline of manufacturing by the government and its excessive concentration on the development of service sector industries (*Guardian* 14/10/86). Labour had initially considered disbanding the Development Corporations but relented, favouring instead the Liberal approach, democratising the Boards and assimilating the frameworks into regional development agencies.

While the UDC initiative became increasingly politicised on the part of central government, it appeared to be less of an issue for the Labour Party at Westminster. This suggests that a vacuum existed in the Parliamentary Labour Party about political solutions to the problem of economic restructuring. Unlike the Government it had no clear response to the massive reformulation of the international economy and the state's part in this process. Additionally the internal divisions in the Labour Party further exacerbated the incoherence over the issue.

Opposition from Labour dominated local authorities was more vocal. This was both politically and territorially inspired. Politically it can be explained by the traditional distance between the Parliamentary Labour Party and local authorities and by the internal political struggle which pitted urban left authorities against the party leadership. The leadership, from 1983, was at pains to distance itself from the radical authorities which had become notorious principally because of their equal opportunities policies (Wainright 1989). Therefore, although the economic strategies were not themselves the object of disapproval, the association between these and other left strategies left local politicians without a clear national framework for the political battle over the approach to local economic development. The dissonance between national and local spokespersons in the Labour Party diminished much of the formal oppositional heat aimed at the UDC, so much so that by 1990, the then Shadow Secretary of State for the Environment, Brian Gould, could be sanguine about the future:

> 'In most cases a simple transfer of UDCs infrastructure and decision-making powers to the local authority would be the best answer' (*Estate Times* July 1990).

Secondly however local responses to the announcement of Corporations, were intensely territorial but varied between and within regions, this only emphasised the uneven development of left politics within England's regions. Even where there was initial opposition it often dissipated and was replaced by a series of collaborative working arrangements.

In Trafford Park there was little disagreement about strategy but much more discussion about the consultation and negotiation structures (*Interview* 9). Nevin has documented the antipathy of the Black Country politicians to the imposition of the Development Corporations, yet they quickly fell into a pragmatic alliance with the Corporation in order to advance pre-prepared strategies (Nevin 1991). Teesside was driven by a fragmented response from local agencies (INLOGOV 1989). The County Council was hostile, and although the borough's responses were mixed, they clearly felt much less antagonistic towards the UDC. In Tyne and Wear the Chairman of the District Labour Party stated a little later that:

> 'the Labour Party was against the UDCs because they weren't felt to be accountable...but we had a choice. Did we stand back and allow the developments to take place and millions of pounds to be spent without Labour Party say?' (*Evening Chronicle* 8/10/86).

Although the arrival of Corporations was generally unwelcome, active opposition was muted, and often parochial. This spawned a complex negotiating arena, local authorities were piqued at the idea that someone else was doing their job:

> 'we thought we could do the job better than they could' (*Evening Chronicle* 7/8/88).

The third generation of UDCs kindled the most trenchant opposition to the overall policy. John Cunningham spoke on behalf of the national Labour Party reiterating a rhetoric shaped by the exclusion of the Labour Party from a central position in state decision making:

> 'The government's policy of extending Urban Development Corporations by imposing decisions on elected administrations of cities and town without agreement or even proper discussion is dangerous. It is symptomatic of the creeping corporate state' (*Daily Telegraph* 8/12/87).

Jeremy Beecham, then Vice Chairman of the AMA, greeted their announcement with the words:

'What we see today is another dose of piecemeal, patchwork policies that marginalise the involvement of people in the inner cities, in the decisions that affect their lives' (*Daily Telegraph* 8/12/87).

Nevertheless the Labour party was unable to maintain a coherent mobilisation against the UDC policy. In this phase both territorial and political considerations became extremely important as some of the new urban left flagship authorities were amongst those which had been targeted for UDC designation. A penumbra of responses sullied forth. In Leeds, for instance, the Chamber of Commerce said that it didn't need a UDC. The local authorities in Leeds had already established a Development Company at arm's length from the council and were rigorously pursuing regeneration partnerships (Roberts and Whitney 1991). Sheffield's private sector differed in emphasis from the local authority, the President of the Chamber of Commerce said that he preferred:

'working within the rules to the imposition of the UDC but I recognise that if the cash cannot be obtained from local government, central government sources will have to be obtained' (*Daily Telegraph* 8/12/87).

But Sheffield City Council was unhappy:

'about styles or ways of working. There were issues relating to social regeneration, and democracy was clearly in question' (*Daily Telegraph* 8/12/87).

Sheffield had established a proactive partnership and had prepared a £120 million regeneration strategy for the Lower Don Valley (Sheffield City Council 1987). This was preciously guarded during the negotiations over the Development Corporation, but so too was the issue of accountability. Sheffield City Council negotiated a progressive Code of Practice with the Corporation, and later entered into an agency agreement with the UDC. Having secured the Code of Practice, Sheffield decided against petitioning Parliament to withdraw the Corporation (Sheffield City Council 1987).

In practice, the opposition to UDC imposition became more sophisticated. It becomes more necessary to separate the political rhetoric infusing the agenda from the actions of officers, civil servants and politicians, involved in that same process.

Manchester concluded similar negotiations to those undertaken at Sheffield, under the influence of Council Leader Graham Stringer. Subsequently a successful bid was made by the council's planning and research unit to monitor the UDCs activities (*Interview* 2). While Manchester objected to councillors being sidelined, the decision not to petition apparently consolidated the transformation in the stance of urban left councils towards UDCs.

Bristol bucked the trend and took their case to the House of Lords.

The reaction to the imposition of UDCs varied, but much of the debate around the third announcement centred on the questions of access and consultation. The politicisation of the negotiating arena depended on the general degree of consensus in the area about the type of regeneration that was favoured. It also depended on the configuration of politics in the locality. The political response to the Corporations did not always come from expected sources. Although there was clearly a central/local dimension to the debates which opened up around the Corporations, the varied responses point to a more complex picture. An element in the Conservative Party and members of the Labour Party were admittedly locked in dispute over the direction of urban policy; but both parties were also internally divided on the issue. The introduction of the 'mini' Corporations saw the 'new urban left' of the Labour Party developing two types of response; outright confrontation such as at Bristol and sophisticated new working arrangement and negotiations, such as those which developed in Manchester. Both of these responses sought to determine the local political agenda, and they achieved very different results: it is this political and policy environment that I will examine next.

The Conservative Party and the UDCs

When Secretary of State Heseltine was criticised for spending too much time in Liverpool, he said

> 'I wanted to understand how these things worked....and to create models...I very much saw the exercise as the basis from which a considerable expansion of the programme could take place' (Heseltine 1990).

It is this sense of experimentation which characterises the whole UDC initiative and links it to the progressive insinuation of the neo-liberal project within the Conservative Party. Jessop et al. describe the three main stages in the strategic development of Thatcherism, as running from the beginning of the Thatcher leadership until the election, followed by a period of consolidation which stretches from the election until 1982 and a final period from 1982 when Thatcher consolidated her control and her programme and deployed them to 'pursue a long term transformation in British society' (Jessop et al. 1988). This is broadly consistent with other accounts (Hall and Jaques (ed) 1989 and Young 1989). Young and Jessop converge in their description of the evolution of the government's strategy as it moves towards the consolidation of Thatcherism. Jessop describes the paradigmatic shift in 1984:

> 'The new accumulation strategy was announced in 1984 (in Lawson's Mais Lecture) and the definition of a hegemonic project. In place of jobs for all and social democracy, Thatcherism would offer the entrepreneurial society and popular capitalism' (Jessop et al. 1988).

Though Young notes the public avowal of a new stance it does not occur until 1986:

> 'A coherent radical momentum would be ...cautiously unveiled at the 1986 party conference. The Next Move Forward...and if resurgent Thatcherism had a theme it would be connected to the institutions ....Now it was the turn of that impenetrable tundra called the Welfare state' (Young 1989).

It is clear from my own description that the Development Corporations emerged as a more coherent, politically charged model of regeneration on the announcement of their second phase in 1986. This was precisely the time at which the 'Thatcherites' had consolidated their control over internal oppositions within the party of government. This political victory initiated a new confidence in the completion of a radical social and economic project. This theme is echoed across a number of documentary accounts and underlines the strengthening of the Thatcherite vision of the state. Yet this clarification of vision is not inconsistent with the earlier characteristics of the government's proposals. After all as Jessop states:

> 'The reorganisation of the state under the past Thatcher governments has been marked by permanent improvisation, trial and error and institutional Darwinism' (Jessop et al. 1988).

The fractious relationship with local government over the search for a new regeneration policy undoubtedly led to a crystallisation in the Thatcher regime, the search for a radical institutional reformulation. The baseline was always transparent:

> 'No government has been elected whose leader was as deeply seized as this one of the need to overturn the power and presumption of the continuing government or civil service, to challenge the orthodoxies, cut down its size, reject its assumptions... and teach it a lesson in political control' (Young 1989).

Nor was it inconsistent with Heseltine's absence or presence in the government. Young describes Heseltine's defence of the inner cities in his early years and his interventionist stance in trade policy. But he also argues that the entrepreneurial swagger of the Secretary of State coexisted with his corporatist tendencies. Despite his apparent contradictions, the Secretary of State was not above shifting the centres of power.

> 'when we created the first Urban Development Corporation I was responsible....for creating new centres of power to do what clearly needed doing' (Young 1989).

In fact, Heseltine's and Thatcher's views on Corporation policy had much in common in the early years as he tells us in his biographical account of the decade:

> 'The loony left in the town halls were a headache for Labour. For a Conservative government which believes in redistributing power away from Whitehall it creates a quandary (in the event the choice out of four available is to) push power out beyond local government and into the hands of people whom it is elected to serve' (Heseltine 1990).

Ultimately it was the late Nicholas Ridley, who was responsible for consolidating the Urban Development Corporation policy. His enthusiasm was fired by three principal factors: the beginning of the boom in the development market, the consolidation of Thatcherism as the party moved towards the 1987 election and his hatred of local government fuelled by the intensity of the battles in the mid-eighties. By the time the second generation of UDCs had been announced the direction of the Thatcherite institutional project was firmly established (Jessop et al. 1989).

In the context of this the UDCs were an experiment in undemocratic processes directed at the locality but controlled by the centre – and ultimately they were about both testing the limits of democracy and thus the limits of legitimacy which the Government could invoke in pursuit of its neo-liberal agenda and the strategic positioning of neo-liberal interests in localities. By 1987 the Thatcherite project entered a period of consolidation. This occurred within her party and without. Opposition was impotent in the face of an increasingly coherent vision of the state economy and civil society. Within this vision the UDCs performed a strategic function. They symbolised the political victory over the left, whether this be in local government or the residual networks belonging to an even more radical yet disparate collage of social and political movements. And they were a flagship of the reorganisation of the state, epitomising a palatable neo-liberal device of state intervention in the local economy and legitimating both this and the effective centralisation of control over urban policy in the locality under a mask of devolution and decentralisation. There was no clearer vision than that which was held by the Thatcherite section of the party, especially when it came to the reorganisation of the public domain and its interaction in the economy. The tentative experiments in state organisation reform now gave way to a fully articulated programme.

The Politics of Implementation: The Development of Strategy in the Local State

By the end of the decade, within the Conservative Party, the policy initiative was consolidated politically and administratively. At the site of implementation something far more complicated and multifaceted was taking place. Many of the Urban Development Corporations met with initial resistance upon designation. Only two of them, however, petitioned the House of Lords against the establishment of a Corporation: London in 1981, (Brownill 1990), and in 1988, Bristol, which were partially successful in their application to the House of Lords. Two other local authorities considered petitioning as a means of resisting the imposition, Sheffield and Central Manchester, but instead pursued the negotiation of complex legal agreements which attempted to secure the implementation of a number of council priorities through the Urban Development Corporation. Apart from these, the Black Country authorities were initially hostile, obtained Queens Council, and considered petitioning, until they were faced with a *fait accompli* by the regional DOE which pointed to criticism of their own strategies and programmes published by the Audit Commission (Nevin 1991).

Interagency Agreements: Negotiating Central/ Local and Private /Public Cleavages

One of the most interesting characteristics to emanate from the UDC experiment is the emergence of an extensive negotiating arena between the local and central and public and private state actors that now constitutes the principal policy making implementation sector at the nexus of local central state relations. Local authorities developed a number of responses to the UDCs, differentiated not only by political considerations but by spatial and temporal factors.

There was some intensive internal disagreement within predominantly ruling Labour groups in local authorities about how exactly to deal with the Corporations. Public posturing occurred which appeared to be about staking a political and a territorial claim that could be useful in future negotiating positions. In addition local authorities often waited for the UDC to make the first move in erecting a framework for operations. Some authorities invested a good deal of time and resources in developing a political and strategic view about how they would deal with primarily political demands. The actual institutional framework on which the relationship between local authorities and Development Corporations was characterised by a series of agreements:

Codes of Consultation – are obligatory in law. They provide the framework for consultation about the strategic and operational work of the Development

Corporation. They have been agreed in the majority of cases, though Brownill reports the ten year period which it took LDDC to develop such a document (Brownill 1990). Most Corporations have included consultation procedures for local communities, primarily about area frameworks and individual projects within the Codes of Consultation.

Codes of Practice – are in effect political agreements between local agencies and the UDCs. Here, local authorities have negotiated with UDCs over elements of the UDC policy and practice. Local authorities have used them to develop a more political agenda. The Code of Practice formulated by Manchester and Sheffield, and attempted by Bristol, contained some of the elements of the Memorandum of Agreement reached between Newham and LDDC after the 1987 election, though the concentration on democratic function was much greater. Sheffield City Council seconded a senior officer to assist in the liaison function from the Corporation. Although Manchester's was based on the Sheffield agreement, it added a number of its own demands.

Agency Agreements – authorise direct handling of planning applications by local authorities on behalf of the Development Corporation. Most authorities have developed these, though Coulson notes the exception of Teesside. As I stated earlier, the small staffs of UDCs necessitated contracting out Development Control work, and Development Control expertise in most areas, lie within the Planning Departments of local authorities.

Joint Working Arrangements – cover projects which are joint funded or developed collaboratively. After their initial confrontational stance, Sandwell and Walsall MBC developed comprehensive joint working arrangements which meant that by 1989 79% of UDC design and implementation work was done by the local authorities (Nevin 1991). Again, Teesside provides an example of a more hesitant local authority, which Coulson attributes, in part, to the style of the Chief Executive who 'believes that the fundamental requirements of economic rejuvenation, namely political certainty and decisive decision-making to match the requirements of incoming industry are irrevocably lost' (Coulson, INLOGOV 1989). Merseyside Development Corporation developed close working relationships as well as Agency agreements with local authorities, especially in the development of a European strategy and, like most authorities, negotiated complex and integrated multi-agency networks of officers. Most UDCs now co-ordinate day to day activity with their counterparts in local authorities.

Memorandum of Agreement – This is a formal document, a contract covering reciprocal arrangements for joint projects and joint funding of projects. What effectively was happening through these negotiations is that the formal rules of a very new game were being negotiated. These were being developed by both right and left authorities. Access to the authorities was to a large extent dependent on the negotiating prowess of the indigenous networks and to some extent on historical associations within the regions. The negotiating arena thus became the new formal 'informal' negotiating structure between the local and the central state. As such it was eminently flexible, depending on the skill of negotiating officers in the locality and the ability of the central state to block further incursions into their areas of their control. As a consequence of this there very different things happened in different places.

Strategic Development

Within this overall framework, localities have also engaged in less formal negotiations with UDCs over strategic developments. These debates have been about the emerging built environment and the distributional issues arising from UDCs. Although some of these arguments have been obscured by cultural and aesthetic concerns, they are in fact about the regulation of collective consumption and the social reproduction of labour power (Harvey 1985a). Although there have been different emphases in the discussions which have dominated local agendas, there are also significant parallels, either in the agreements that have been reached or in the controversy that has been generated.

Most UDCs have experienced some conflict with local authorities over *housing*. While most have been grounded in equity concerns, i.e. the target market or the merits of residential versus job creating manufacturing development, there is also some divergence in the way that this issue has evolved. Liverpool in particular saw the emergence of new organisational and political alliances as a result of the powers invested in the UDC. Here a clash arose between the municipalist housing policies of Liverpool City Council and a local community group, the Eldonians, who wanted to develop diversification in tenure and who participated in a UDC campaign to extend the boundaries. The Black Country Council and the Development Corporation differed on their strategic goals for housing especially in relation to environmental improvements (Nevin 1991). Social housing was also an issue in Bristol, where no social units have been built, (*Interview* 5) but Tyne and Wear responded to the LDDC situation by focusing on social housing as an asset in negotiations with the local authority (TWDC 1989/90).

The second source of major disagreement between UDCs and Local authorities and one which has repeatedly occurred, concerns *retail* development. This echoes a broader concern which the new spatial patterns of retail development were presenting to local authorities and established retail traders. Although Meadowhall in Sheffield was initiated prior to 1987, it caused local tension as retail rents fell in the town centre and it transpired that the jobs in Meadowhall were largely relocations from elsewhere in the metropolitan area (Dabinett and Ramsden 1991). Councils with and without UDCs raised fears about damage to established shopping centres by peripheral development throughout the 1980's.

The difficulties which Corporations have experienced over *infrastructure* have often involved relocations of existing businesses or in the case of roads in particular, where the development threatened environmental disturbance or damage. Infrastructure development has therefore often galvanised opposition from small businesses as well as environmental groups. This has taken place in Sheffield and in Bristol. Similarly, arguments have raged over the two barrages, the first at Teesside and the second in Cardiff. In both cases political opposition has been split.

On *employment*, demands from localities have centred on local training and employment compacts. In fact the mounting criticism over employment generation escalated after the LDDC debacle, and the publication of the 1988 Select Committee on Employment. Liverpool established a positive reputation around training in its early years when its training figures compared favourably with those of Dockland's. But in the late eighties apart from Teesside, all of the second generation Corporations established a high profile on training. Sheffield and Trafford Park in particular have been keen to publicise their training agreements and all have developed close working relationships with their local Training and Enterprise Councils, especially as they often have board members in common. The third generation of UDCs have been less proactive in the development of training packages.

Much of the open debate between the UDCs and local agencies centres on access and democracy. The strategic issues were often then dealt with in the context of formalised semi private working arrangements. The public spats in addition erupted over territorial issues or where groups had been excluded from the semi formal negotiating arena or indeed if the issue of the distribution of benefits remained unresolved.

There has been remarkable little disagreement between authorities and Development Corporations over the strategies of *place marketing*. These have been enthusiastically pursued by all of the UDC areas (with perhaps the exception of LDDC). They appear to be seen as universally 'good' strategies by all of the local agencies and partnerships.

The underlying contours of strategic disagreement have been rooted in ideological, territorial or functional interests. Many of them fall into the latter two categories, for instance the future viability of city centre retail and commercial development versus out of town shopping mall development, or whether a road is a good strategy or not. But much of the unease that is ideologically inspired is rooted in a dispute over the meaning and intention of regeneration and the way in which regeneration should be carried out and by whom.

There are two features of the development of UDC planning frameworks (as distinct from overall operational frameworks discussed above) which have been particularly important. Existing plans for local areas were often thrown out by the UDCs which developed entirely new frameworks within their vision documents. In addition the style of implementing planning policy which, in the urban left authorities in particular, emphasised consultation and participation by local residents has also been important.

There seem to have been three responses to the first issue. The first response was outright confrontation such as that which took place at LDDC. Here not only did the new agency contravene the existing district and statutory plans, local authorities then used the previous organisational frameworks as the basis of resistance for to the new plan. Most UDCs remain a black hole in the middle of a Structure or Unitary Development Plan. Secondly, local authorities and local groups have incorporated through struggle or negotiation, their own structures and ideas into previously existing planning frameworks, such as in the Kirkstall Valley area in Leeds or the Don Valley Forum in Sheffield or indeed the Black Country where Nevin reports that:

> 'it could be concluded that in its initial stages the Development Corporation utilised the management and strategic experiences of the local authority'. (Nevin 1991).

Kirkstall valley is one example where intensive 'planning for real' exercises were seen as part of a dynamic approach to development proposals.

Finally, some Development Corporations simply inherited the planning framework, did not disagree with it and, after small modifications, proceeded to implementation. Even where this has been the case however there have usually been some difficulties with flagship development. Almost all of the Development Corporations use flagship developments. Although most UDCs have combined this approach with some level of strategic planning, there have been exceptions, such as at Teesside. Since LDDC, almost no UDCs have advocated the complete abandonment of planning frameworks.

Almost all rely on area planning frameworks, which accommodate a market-led approach, though often they also provide for 'public consultation' centred on public meetings and/or referenda or by mailing their plans to local residents and businesses. UDCs have encountered a variety of responses to their strategies, though in reality the difference between the Corporations' plans and those of the preceding local authorities has often been insignificant. In planning style, the most observable disputes have been about public consultation and the level of integration with the strategic dimensions of Structure and Development

Plans. On the issue of organisational style, dissatisfaction has been about the emphasis on a market led approach, or the degree of secrecy surrounding the decision making procedures of the UDC. Although different planning and organisational issues have come to the fore across different regions there has been significant variation in the negotiating stance adopted in each of the areas.

Political Opposition to the UDCs

The strategic opposition to UDCs, which became evident in the third round of designations was largely exogenous to the locality. As the number of UDCs had increased, so opposition to them began to crystallise in the formation of a series of national networks. While the networks had links with the Labour Party at local and national level, they were not party political. Rather they comprised activists from local authorities, voluntary organisations, political movements and academia who sought to develop a more critical approach to the Corporations. The pre-eminent organisations within this arena, were CLES, based in Manchester, and INLOGOV, part of Birmingham University. Their organisational foci and their political objectives were slightly different, although a number of local authority officers, as well as the Docklands Consultative Committee, and the National Council for Voluntary Organisations belonged to both of them.

CLES is a subscriber organisation for local authorities and is committed to the popularisation of alternative economic strategies (which emanated from the most prominent new left authorities during the 1980's). INLOGOV on the other hand is an academic organisation concerned with a broader analysis of local government and a commitment to promoting good practice in local government.

Both networks built upon the criticisms made by the 1988 Select Committee on the employment effects of Urban Development Corporations (Employment Select Committee Third report. HOC 327-1. 1987-88) and the later NAO Report on Development Corporations (NAO 1989). CLES in particular established a critique linking the unaccountable nature of the Corporations with the lack of social equity and redistribution of benefit that has been associated with Corporation activity. CLES contrasted this approach unfavourably against that of many inner city local authorities. The comparison is not strictly fair, in that many local governments have eschewed an equity led approach to urban regeneration. In fact local authorities display as much variation in the implementation of urban regeneration policy as UDCs. INLOGOV was more conciliatory and appealed to a broader cross section of local authorities. They established a database and commenced regular monitoring of the activity of UDCs. Although some objectives were common to both networks and there was some dual membership, INLOGOV's approach was less politically contentious. Both organisations have published independent monitoring reports on Corporation activities.

INLOGOV reported first in 1990. They provided information on unemployment change between 1988-90, sectoral employment trends within the Corporations, breaking down gross employment figures into construction, manufacturing and service industries. They also analysed land acquisition in categories of vesting, purchased and compulsory purchased, giving a broad, though incomplete picture of where difficulties in the acquisition of land might be taking place (INLOGOV 1988-90). Other information, such as new building in housing, industrial and commercial floor space and the planned public expenditure for UDCs, would, by 1992, be contained in DOE Annual Reports. Thus INLOGOV concentrated on evaluating the contents of UDC plans. CLES also produced a monitoring report in 1990. This was highly critical of UDCs vis a vis social equity issues. It provoked stinging responses from some of the UDCs, including TWDC and Trafford Park Development Corporation (*Interviews* 12) to the extent that some of the Development Corporations will no longer speak to CLES researchers. CLES however was interested in the issues of process form, particularly those related to accountability.

The networks were important for a number of reasons. In the light of the failure of political parties to organise a resistance to the UDC, the networks consolidated a series of substantive and political initiatives across party political and traditional craft orientated trade union boundaries. The networks served as a surrogate opposition to the government and challenged the objectives, to some extent the rules and the nature of local government's dialogue with the UDC. They also provided communication channels between regions drawing in statutory and non statutory organisations. They were thus able to influence some of the more interesting experiments in negotiating leverage with UDCs on both the content and the process of UDC policy. Finally they were able to maintain a semblance of commitment on the part of some authorities to the promotion of alternative economic models of regeneration. In effect they served to politicise the struggle between UDCs as representatives of working class interests across a national arena.

Conclusion

As the decade closed, UDCs were placed within an elaborate framework of influences. A myriad of institutions complemented their efforts to support the elevation of city regions above the annihilation visited on local economies. On the left of the spectrum, Enterprise Boards, local employment strategies, experimentation in the collaboration between local authorities and local industrial networks to create alternative socially useful production relating to sector and regional strategies proliferated, especially up to 1986 (Boddy and Fudge 1984). The right suggested the curtailment of local authority economic powers, transferring them to newly established institutions insulated from locally accountable political networks (Dunleavy 1991). A penumbra of initiatives including many of the local authority or private sector regeneration partnerships were established somewhere in the middle of the ideological continuum. The competition for ideological supremacy and public legitimacy was evident as the models rose and fell throughout the 1980's and as national and local forces developed new tactics in overcoming resistance to successive innovations.

It appears that political opposition to UDCs at a macro level gave way to a series of negotiations between Corporations and local authorities. At the height of the political row between central and local government in the mid 1980s, the plans of the left and right were polarised (in the most extreme cases) on the basis of a central local/left-wing/right-wing division split over the strategies which were being pursued and the methods of implementation. After 1986 this situation changed and we have observed since then a more elaborate conflux of tactics and positions. Political, territorial and strategic struggles have taken place primarily in the locale. Undoubtedly they are facilitated and mediated by national networks. The struggle over process is clear in the negotiations about the nature of consultation and working relationships with local authorities. The struggle over strategies has often been located within a national framework of urban policy. It is about the question of who benefits from urban regeneration. One of the most interesting developments to come out of the implementation phase of UDC policy is the growth in a diverse range of agreements which to some extent appear to indicate the political environment in which they are drawn up.

Within the organisational and planning framework of the UDC, planning gain became a key negotiating concept in securing local benefit. As relationships developed between the UDCs and local authorities, so the complexity of operating agreements increased. It seems clear that these reflect the tensions between the different ideological trajectories of state policy. In contrast to the proposed organisational and philosophical framework UDCs have, on the whole, adopted a strategic approach. The plans developed by UDCs however often co-exist as a black hole within Structure or Unitary Development planning frameworks. There are substantial differences between the generations of UDCs, much of which we have accounted for by referring to the gradual consolidation of the Thatcherite hegemony, the influence of national networks and the results of the initial experiences of UDCs shaping the activities of later generations. But there are also significant differences across the UDCs which were established in the same period. This can be accounted for by spatial peculiarities and the complexion of different political constituencies.

It seems clear however that UDCs were established as a hegemonic vehicle, not so much for what they were to do, but how they were to do it. The implementation process was not as smooth as had initially been anticipated. Ostensibly the UDCs were about creating new inner city constituencies, establishing state interventionism in local economic development as a private domain and challenging both the power and the ideological thrust of the new urban left; their success, I will go on to argue, has been somewhat mixed.

Notes

1. Though this policy has altered somewhat as Brownhill illustrates in her book Developing London's Docklands: Another Great Planning Disaster (1990) PCP. London.
2. As input/output measures the indicators have not really measured process of developing Corporation policy and actions. The only internal 'process' report to take place during the first eight years of the implementation of the policy was one conducted by the NAO. This looked at the first two Development Corporations – LDDC and MDC – and conducted a Financial Management and Policy Review. The review concluded that the Corporations 'had placed too little emphasis on setting precise priorities and on financial planning'. In fact the review also advocated a greater emphasis upon the strategic element within Corporate plans and upon monitoring and evaluation. Subsequently the NAO have conducted another review of the UDCs. This reported in 1993, but again the concentration was on outputs. Manchester University is currently preparing the final review of the overall policy which will take place on the 'exit' of the Corporations from the urban policy arena.

2 A framework for the analysis of the Urban Development Corporation

Introduction

This book attempts to provide an empirical and theoretical explanation of the likely metamorphoses in the state form, as society progresses from one economic and social paradigm to another. It is my contention that the level and pace of change, not to mention the relative success and failure that we can attribute to the transformations that are taking place, are enormously influenced by the configuration of state, polity and economic relations within urban and regional spaces. This is in itself, hardly a contentious assertion, it is almost a truism to say that one area of regional discourse will inevitably influence another part of its development. However much of the academic and theoretical dispute about the dimensions of the linkages and their cross sectoral impact, is rooted in the precursory question underpinning the book. Are we actually progressing from one paradigm to another? The answer to this question lies, it seems to me, in another question. How can we tell if we are progressing from one paradigm to another? It is this attempt to show how we might possibly be moving from one paradigm to another, which preoccupies this chapter. It will suggest that a key unit of analysis in observing such change is the policy processes of the state, as they operate at the local central nexus in the urban regions.

I want to develop an interpretation of 'process form' within contemporary state institutions that will link the development of policy styles and frameworks developed in the UDC, to the macro processes of economic, technological and organizational evolutions in the locale. There is no doubt that the policies and practices that are evolving in the Urban Development Corporation are symptomatic of a dramatic change in state form but does this mean, as Harvey suggests, that we are experiencing a struggle over the creation of a new 'mode of social regulation' (Harvey 1989a)?

My model sees the state as a site for the resolution of antagonistic relationships in the interest of capital accumulation. Simultaneously the processes, which carry and mediate the relationships embedded in this arena, are dependent on the apparatus of the state to maintain their legitimacy and their operational continuity. Without these mutually dependent formulations, the stabilisation of the capital accumulation process and of a functioning civil society would be difficult to maintain.

I argue in the following pages that the missing link thus far in analyzing the role of the UDC, within the context of changing regulatory paradigms for capitalist society, is David Harvey's conceptualization of 'structured coherence' and 'urban class alliance'.

In order to unpack the linkages between one level of analysis and another, that is, the nation state, regional development and local policy processes, I want to go on to consider some of the fundamental concepts with which one can construct a framework to analyze state transformation and the motives impelling change.

Class

Any understanding of the dynamics of state and regional relationships and their periodic transformation, should incorporate an understanding of class and class activity. This is because it would appear that the dynamics of change in the state and urban region, are rooted in that now old fashioned concept of the struggle between classes. One Marxist tradition is to locate class conflict solely in the realm of production. Others however, including Harvey see class conflict resonating through the realm of circulation and exchange. In the Limits to Capital (Harvey 1982), Harvey explores the nature of class relations and identifies two dimensions of the concept of class: 'class as it relates to the mode of production', that is the relationship between capital and labour as it is manifested over the rights of labour power, and class as 'social formation'. This second categorisation of class is a historical concept (as opposed to an economic relation) and it embodies a range of class relations embedded in production, exchange, distribution and consumption, religious) institutional and juridical and administrative organisation. The concept of class as a social formation, is derived from Marx's writing in the Eighteenth Brumaire of Louis Bonaparte (Marx 1978).

It must be said that there is no systematic view of class within the works of Marx, and I have deliberately chosen the interpretation which he arrived at in the Eighteenth Brumaire. This accords entirely with my own view about the complexity of class formation and fractionalisation within late modernity, as well those authors such as Overbeek and Jessop, who are similarly grappling with the problematic of class identification in a post Keynesian state (Overbeek 1990, Jessop 1990). The Eighteenth Brumaire of Louis Bonaparte conceives a multiplicity of classes within capitalist society. As well as two major classes, the proletariat and the bourgeoisie, Marx identifies transitional and a number of middle classes. For Marx, the very existence of classes meant that class conflict was inevitable, and indeed it was this conflict which was the motor of history.

Struggle therefore becomes an engine of capitalist development. Collective and thus, class struggle occurs, as a result of the incentives to competition and the production of surplus, on the part of individual capitalists. This results in the necessity of further degrees of 'exploitation'. But competition does not simply take place between capitalists; it also divides labour. Without any collective combination of labour as a class, the exploited have no individual resource or power to withstand the onslaught of capitalists. And 'as labourers offer collective forms of resistance, the more capitalists are forced to constitute themselves as a class to ensure collectively that the conditions for progressive accumulation are preserved' (Harvey 1982). In the Eighteenth Brumaire Marx distinguishes however between 'class in itself' and 'class for itself'. The working class is a class in itself because of its position within the system of production in society. It can only become a 'class for itself' by becoming conscious of its collective common interests and through its collective action (Edgell 1993).

According to this interpretation class consciousness becomes crucial in the evolution of class struggle. Consciousness would appear, within capitalist society to be specifically about realising the collective potentials of exploitation and resistance to exploitation. Wright has defined exploitation as the 'economically oppressive appropriation of the fruits of labour of one class by another, (Wright 1989). He emphasises, importantly in the context of my own analysis, that it is not only capitalists who are able to exploit non owners on the basis of their organisational assets and/or their skill credential assets. The institutional capacity to nurture collective consciousness is an extremely important dynamic in the power relations between and within classes. In late modernity, the ability to overcome the multiplicity of cleavages and contradictory interests within society makes this an almost impossible task. Because of this, the issue of class fractionalisation and the mechanisms for uniting or dissipating fractionalisation, assumes greater importance within contemporary analysis of conflict and consensus between classes. So what then is the impact of fractionalisation, and how does it impact upon state relations?

Overbeek conceives of the capitalist class as five fractions, focused on the different 'functions', 'positions' or 'moments' within circuits of capital. Fractions of capital are seen as: 'groupings of capitalists with structurally compatible positions within the overall circuit of capital'. Conflict between fractions tends to transcend competition between individual capitalists' (Overbeek 1990). Fractional interests have, in Overbeek's view, a structured economic function, which, in the contemporary economy, divides into a production function where industrial capital is dominant, and where the workers are an essential element to be considered when there is a conflict between the classes. And the circulation function in which money capital (that is banks, building societies and insurance companies) is dominant and where an essential element to be considered when a conflict in the classes occurs is the movement of the money form, so, in this instance, white-collar professionals, homeowners and investors are the essential element.

Classes can no longer be seen as simply capital and labour. They have become tied up with multiple functions of capital. Therefore working class interests in the production function rely on the continuity of production and the sale of commodities. In the circulation function however, working class interests are more concerned with guarding the free movement of capital, the largest freedom of manoeuvre and the minimum of regulation, this tends to be in the professional financial sector.

Having examined the spatial and functional elements of class, I want to move on to look at the possibility of contextualising class within the state by making use of Overbeek's 'concept of control'. It is this notion which draws us closer to the struggle that is taking place in UDCs, and provides a theoretical connection with the arena of 'urban class alliances'. Overbeek uses a 'concept of control' as the key factor in creating the precondition for class consciousness and facilitating collective class action overriding individual sectional interests (Overbeek 1990). The 'concept of control', can be translated as a 'hegemonic project'. If a dominant class fraction can unite other fractions under the guise of acting in the 'general interest' they have successfully united the capitalist class in a strategic action.

Overbeek argues that there have been four concepts of control since the 19th century, classic liberalism, state monopoly tendency in the early 20th century, corporate liberalism 1940 to 1970, and the current period is marked by a neo-liberal concept of control (Overbeek 1990). The hegemonic project becomes supremely important to the establishment of class power over the rights of exploitation. Poulantzas had previously developed this idea:

> 'The state organises and utilises divergent interests of different fractions of capital under the hegemony of the dominant fraction' 'we do not find wage labour and capital necessarily confronting each other directly at a political level, but this does not mean that groups do not confront each other and are not engaged in class struggle' (Poulantzas 1975).

Therefore the strategies which comprise the neo-liberal project, arguably, determined in the present context by the dominant faction in the Conservative Government, can be seen as a part of a hegemonic project. This is designed to unite fractions of capital in the United Kingdom, to act in the interests, it is argued by Overbeek and echoed here, of finance capital. But such a strategy is likely to be contested. Not only by interests within capital, but by elements of labour and conceivably by elements still rooted in the state monopoly tendency. Within this framework, it seems to me that the Urban Development Corporation could be seen as an instrument in a hegemonic struggle to unite class fractions in a collective consciousness committed to a particular class project, that is the change from one economic and social paradigm to another.

Latterly, Wright has utilised the notion of 'contradictory locations' to identify the growing and diverse intermediate sector within late modernity. His goal is the development of an adequate marxist class map, as yet unachieved. In Wright's second class map he divides owners and non owners of capital and overlays this with a further subdivision between the state sector and the capitalist sector. This creates twelve class positions in contemporary society, and a number of 'contradictory class locations' (Wright 1989):

> 'Within capitalism, the central contradictory location within exploitation relations is constituted by managers and bureaucrats. They embody a principle of class organisation which is quite distinct from capitalism and which potentially poses an alternative to capitalist relations' (Wright 1989).

Wright concedes that the balance between sectors within the alliance formation depends on 'history cultures and struggle'(Edgell 1993). I would agree with Wright, though my own interpretation departs from his at this point to coincide with that of David Harvey and Anthony Giddens. One could argue that state bureaucrats do indeed occupy contradictory class locations given the multiplicity of forces extant in their development, but equally it is possible that the overall positions of particular bureaucrats in urban regions are profoundly shaped by the dominant forces in an 'urban class alliance'. In other words though these actors themselves shape their environment, the arena in which they operate is subject to a dominant set of interests. All sorts of influences shape the perceptions of these actors and the perpetual struggle in which they are engaged, is to achieve a collective consciousness within the fundamental division of capital and labour. The key struggle between these groups, is to control the mechanisms of the state so that their strategies can be implemented. Control of the mechanisms it is argued elsewhere flows through the rules and resources of state processes (Healey 1990). If there is a continuous struggle over the rules and resources embedded in the state and if this has a connection with the accumulation process, who or what controls these rules and resources becomes a crucial question in determining the dimensions of relationship between the economy and the state.

The debate between agency and structure as determinants of state form has for too long crippled the social sciences (1). Giddens has gone further in upgrading the role of the individual actor in this scenario. He does not see power and structure as separate entities but as a dualism, power and structure for him are interpenetrated. He terms this 'the duality of structure'.

> 'human agency reproduces structure which simultaneously serves as the condition for the reproduction of human agency in a continuing process, human agency is irreducibly tied to power, without power there is no agency' 'in promoting the notion of agency he suggests that' it is the knowledgeability, skill reflectively and receptivity of the human being which requires interpretive meaning' (Giddens 1984).

According to this conceptualization actors are a crucial mechanism in realising the connections and correspondence between the accumulation process and the state form. These actors are reflexive and therefore interpenetrate the state structural discourse with their own knowledge, their cultural and their class positions. The expansion of the Keynesian state enables this dialogue between a range of actors and their structural positions.

Within the context of a a reorganisation of the state form, the potential for creating class consciousness, can be seen as the reorganisation of the capacity of classes to develop a coherent identity or indeed to prevent the combination of class fractions in order to hinder such a process.

It is possible therefore that we are observing attempts in the locality to unite a class around a hegemonic project and a series of counter attacks to bring other interests together. The state bureaucracy is not the only group likely to occupy a contradictory class location within this scenario. Other actors within the social formation may operate within different modes of production, according to Regulationists and their precursors such as Poulantzas:

> 'in every social formation, we find the dominance of one mode of production which produces complex effects of dissolution and conservation on the other modes of production and which gives societies their overall character' (Poulantzas 1975).

Jessop also argues that there are several modes of production within any given historical period but one dominant mode (Jessop 1992). This scenario carries with it the possibility that the many social formations premised on these modes will also be present, as well as the residual social formation from previous historical epochs. Therefore the material foundation of state policy is very much connected to the status of the economic and social formation of the area and the cohesion or disparity in the strategic decision-making structure is equally premised on the nature of interpenetration of modes of production from different historical periods.

I have argued in the previous section on class that it is useful to see the centre as drawing up a 'comprehensive concept of control' in the interests of a particular section of capital. Therefore the hegemonic function of the government in Whitehall and Westminster becomes particularly important in the context of transforming state capital relations. Stoker illustrates the extent to which the centre has been involved in controlling and curtailing the power of locales, but the proliferation of local state forms is not a one way process, it is interactive and responsive. This being so it behoves us to examine the multi-organisational environment in closer detail and more than ever to concentrate on processes as the key mechanism for interorganisational fine tuning.

Structured Coherence

David Harvey is central to my efforts to link capital, state and region because of his contribution to the development of Marxist thought by introducing space into the analysis of social relations. By focussing on the development of the built environment and linking it to the realm of circulation relations and crisis, he has allowed us to suggest the possible linkages between economic and state restructuring. 'Structured coherence' describes the harmonious development of technology, organisations and social relations which ensure successful capital accumulation within the urban region. While placing the global local debate firmly within the logic of capital accumulation process, it also suggests that a whole range of cleavages interact with the competitive processes of capital accumulation to produce unexpected outcomes (Fig. 2.3). Within this conceptualization Harvey parallels some of the propositions of Regulation theory. He shares with Overbeek the broad conception of fractions of capital and labour determining the nature of struggle and the resolution of crisis (Harvey 1985a). Harvey's analysis is shaped by his concern with spatial determinants in the accumulation process, whereas Overbeek conceptualises the fractions within their functional form, intersecting with historical and conjectural forces.

Harvey allows us to analyse the state in a transitional period at a sub national level without falling into the trap of assuming that the new correspondence between global capital and the local state suggests that 'post Fordism' is premised on a move to local regimes of accumulation. The term transitional may not simply indicate a move from one paradigm to another – it may indicate a switch in the dominance of particular circuits of capital (Harvey 1989b). Local regimes of accumulation, I suggest, do not exist outside global circuits of capital. Rather, the role of the urban class alliance gives a material form to the ensemble which

Joe Painter speculatively entitles a local mode of regulation (Painter 1992). A local mode of regulation is of course central to decisions to switch from one circuit of capital to another. One potential element of this 'mode of regulation' is the Urban Development Corporation, which, it is argued here, is a site of struggle over the shape of the emerging class alliance after the present round of restructuring. It allows a conceptualization of the state and local economy relationship that it is not crudely deterministic, but which nevertheless offers a compelling structural argument.

How then does this macro-economic framework relate directly to spatial considerations and what relevance does it have to the concept of class relations and their impact on the UDC? Harvey connects these issues through the mechanism of 'structured coherence'. Structured coherence materially underpins the coalition of actors who make up the regional class alliance. In order to understand this we have to appreciate how space and accumulation are associated in Harvey's framework. Harvey builds upon uneven development as a means of incorporating crisis by exploring the relationship between spatial form and the built environment. He explains that the circulation of capital needs to be completed in 'socially necessary turnover time'. Put very simply, this means the quicker the turnover time for individual capitalists, the greater the profit they make. Harvey reminds us that fixed capital has a turnover rate that is much slower than is normal.

As surpluses of both capital and labour are continually being produced, they are easily shifted into long term infrastructure development, where 'a part of the circulation process slows down in order to promote faster aggregate turnover time' i.e. from primary to secondary circuits. The mechanism through which it is engineered is 'fictitious capital' which, in the form of shares, bonds etc., allows an annual rate of return which mitigates the long drawn out circulation process. But in order to continue this process of 'dynamic equilibrium', it is necessary to promote continual temporal displacement through accelerating fictitious capital formation. This can precipitate a crisis, as increasing amounts of living labour is dispatched to work off past obligations, and begins to manifest the conditions for the falling rate of profit. Inevitably this sparks off a crisis of overaccumulation. More capital is stored in physical and social infrastructure, overaccumulated capital flows back to combine with greater reservoirs of surplus capital and eventually the capacity for more fictitious capital is blocked by various constraints which mean that there has to be a devaluation.

One of the means of coping with this crisis is to displace it and thus the 'spatial fix' becomes an important instrument of crisis management and underpins the concept of uneven capitalist development, that is, a reservoir of spaces and reserves which can be usefully plundered as geographical or sectoral crises occur (Harvey 1985b) The spatial fix, however, can and has to be understood more generally within the concept of time and space relations. Harvey suggests that time is crucial to the production of profit, as socially necessary turnover time. Within this context space is seen as a barrier to achieving socially necessary turnover time. Space as a barrier can be overcome by production in space. Harvey shows how the production of spatial configurations assists capital to 'annihilate space with time' and in doing so becomes the point of yet another contradiction. Spatial relations are thus continuously subject to transformation.

I want to now explain what happens in the space where these forces become conjoined. What happens in this space, is the tendency of the forces towards a 'structured coherence', of organisational form, technological mix and social formation. Harvey defines this as a territory that corresponds to a reasonably coherent labour market, which is later formalised in representation by the state. If Harvey had left his analysis at this point he would have done little more than elaborate the details of capitalist uneven development and crisis. But by developing structured coherence and the urban class alliance, he converges with Regulationists to give a spatio-temporal pattern of the social relations of circulation and exchange. This allows us to grapple with the complexity of class state relations in the urban region.

The Urban Class Alliance: Its Role and Function

Thus we come now to an explanation of how class forces configure in the urban regions and particularly in relation to the state form. An alliance is made up of actors who represent both capital and labour; Harvey does not suggest how the members would be recruited, or who they might be. He does speculate, on how the representatives of labour and capital might perceive their interests in an area. (Fig. 2.2).

Harvey's 'urban class alliance,' is a group of actors who formally and informally manage the economic and social dynamic of a region. They are materially underpinned by 'the tendency for the economy to achieve ...structured coherence defined around a dominant technology of both production and consumption, and a dominant set of class relations' (Fig. 2.3) (Harvey 1985a). The urban region is geographically contiguous with a labour market facilitating daily exchanges and substitutions of labour power. Structured coherence is a present tendency within this labour market. Here a dominant mode of production and consumption exist. The wages of a workforce are converted into a pattern of consumption and a market for goods and services. Although less capital is confined to a particular urban labour market, Harvey suggests that 'the volume and qualities of the internal labour market are important parameters for capital operating there', but the modes and mechanisms of distribution are particularly important in determining the area's level of coherence.

Figure 2.1 Multiple Dimensions of Class

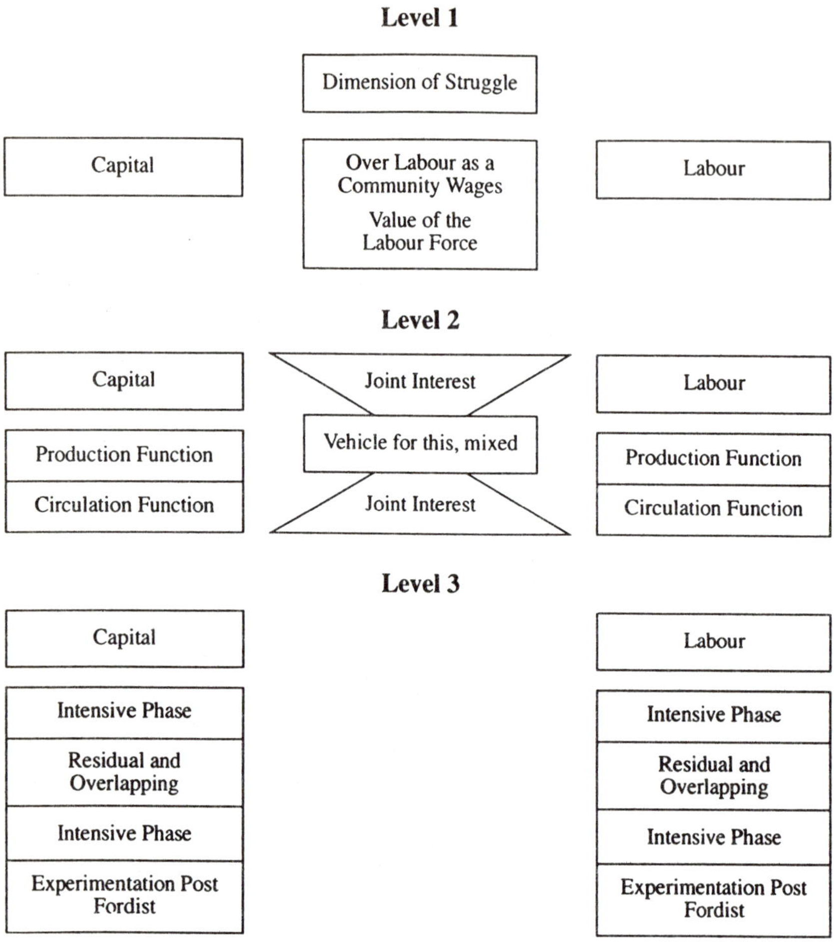

I talked at the outset of the social and physical infrastructure underpinning the structured coherence and the consolidation or dispersion of a coherence as having a lot to do with the dominant and regressive predisposition of the region towards one of three possible circuits of capital (Harvey 1985a). It might be useful to explain this further. The physical infrastructure of cities – roads, schools, factories, shopping malls – exist for both production (fixed assets) and consumption (consumption fund). This capital stock absorbs large amounts of geographically immobile investment. In effect, the urban region is defined as a 'spatial configuration of a built environment for production, consumption and

Figure 2.2 The Ruling Class Alliance

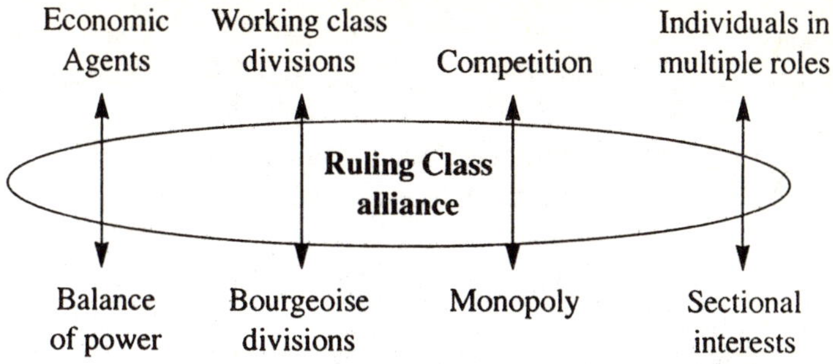

exchange' (Havery 1985a). Harvey suggests that this configuration embodies a dominant technological mix and certain capitalists have an interest in promoting and/or maintaining such infrastructural investments. But in order to invest in the built environment, capital has to be withdrawn from current consumption. This is usually done by debt financing, at the same time capital invested in the built environment is vulnerable to devaluation so the protection of such assets is extremely important for capitalists. Therefore the involvement of the Keynesian Welfare state in terms of both its local and national formations, in the development of the built environment and the movement and switching of capitals across different circuits has profound implications for the nature of the state and the nature of capital accumulation.

Harvey is less specific about social infrastructures, but he suggests that these involve the state institutions:

> 'within the urban region institutions and the people who run them tend to coalesce, rarely without conflict (but) into a matrix of interlocking and interdependent social resources offering a specific mix of social possibilities' (Harvey 1985b).

Overall, despite the social and cultural difference within an area, there are strong material and socio-cultural impulses towards creating a regional identity. The stability of a ruling class alliance and thus the precondition for economic stability or growth in the urban region, is predicated on the stability of the structured coherence, but 'competition, accumulation and technological change' are destabilising factors and it is within this potential for destabilisation that 'a political space with in which a relatively autonomous urban politics, can arise'.

Capitalists who have invested in the locale, for instance, land and property owners, as well as fractions of labour who are homeowners, are likely to be more committed to a region. These groups meld in the interest of protecting their

Figure 2.3 Structured Coherence in the Urban Region

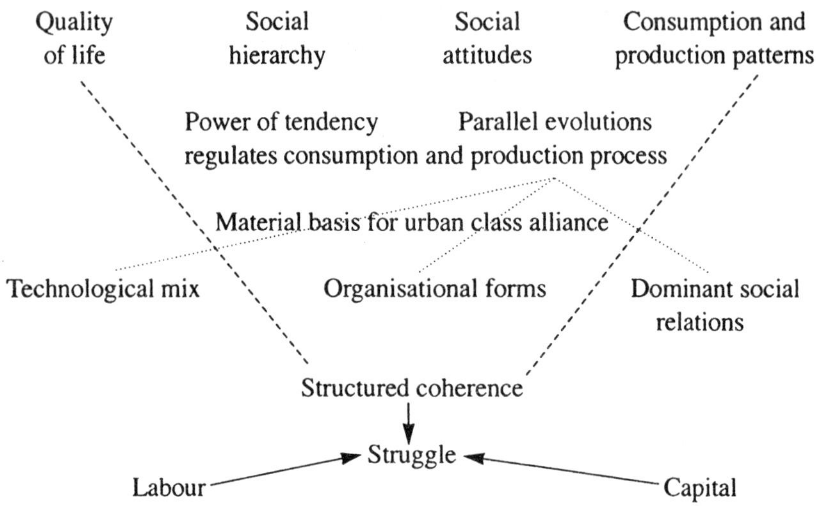

investments and they are often central (as producers, consumers and rentiers) to the decision-making procedures around the development of the built environment (Harvey 1985a). If I take Harvey's analysis further and incorporate the way in which capital progresses through primary, secondary and tertiary circuits in its attempt to secure valourisation, the centrality of an alliance to investment decisions in an urban region becomes even clearer.

As Harvey has explained, manufacturing activity takes place in the primary circuit of capital accumulation. When over-accumulation takes place there is a movement of surplus capital and labour into the secondary circuit i.e. building and infrastructure and thence into the tertiary circuit i.e. education, the production of knowledge etc. Wherever the dominant interests in a region are tied up with one circuit in particular, they will resist movement into other areas. As the built environment is a particularly immobile area, interests tied up in this are crucial.

In addition to this, major economic forces push capital to stay in one region or move to another. This can change the whole nature of the alliance. Similarly, if the production process or the labour market structure alters, this will also cause instability in the alliance.

Nevertheless, fractions within capital, ie financial, commercial, industrial and producer, and within labour, i.e. men, women, white, blue collar, public or private sector, are as likely to be pulling in different directions as they are in the same direction. So how do they provide stability for accumulation?

The criticism which can therefore be levelled at Harvey is that the class alliance comprises 'every one but no one in particular' (Harvey 1985a); in other words it is no better than a pluralist interpretation of how groups coagulate and dissipate around systems of power in local and regional governments. However both Harvey's and Overbeek's arguments about the constitution of groups involved in conflict over the state would suggest that the conflict can be traced to arguments about the direction of strategies of capital accumulation, whether these be production based strategies or consumption based strategies, and that the struggles over access to the decision making structures around these strategies is weighted towards certain strategic class interests (Offe 1985).

In arguing this I refer to two propositions made by Harvey, both of which seek to answer questions around the growth of new social movements and the introduction of identity politics into the urban agenda during the 1970s (Scott 1990). Harvey deals with the issue by looking at the politics of consumption, the struggle here is over the reproduction of the labour force:

> 'we also have to consider the costs of reproduction of labour power at a standard of living which reflects a whole host of cultural, historical, moral and environmental considerations. A change in these costs or in the definition of the standard of living has obvious implications for real wage demands and for the total wage bill of the capitalist class. The size of the internal market formed by the purchasing power of the working class is not irrelevant to accumulation either. Consequently the consumption habits of the workers are of considerable direct and indirect interest to the capitalist class. But we should also consider a whole host of qualitative aspects of labour power....and the variety of fragmentations along racial religious and ethnic lines' (Harvey 1985a).

This leads him to develop the notion of 'displaced class struggle', struggle that may originate in the workplace but that 'reverberates' throughout the realm of circulation relations. This includes a whole host of struggles around not only consumption but around quality of life and environmental issues, issues such as location, community voice, disputes around community and competition.

These struggles can all be seen in the context of their impact on capital flows and investment strategies, and they are capable of making or breaking the flows of capital within and between regions. Thus the narratives which evolve around 'inter-urban competition' can be seen as class narratives.

The way in which the state developed in the post war region and its integral relationships with decisions about the movement of capital, means that a significant proportion of people, and politicised forces, had access to the decision making arena when it was making vital decisions around capital accumulation.

Harvey points to a role for urban politics within the continuously revolutionary politics of capitalism. The constant innovation which is required in production needs parallel innovation in consumption patterns, bureaucratic and administrative regimes. One of the characteristics of urban politics, he tells us, is that it is 'as speculative, innovative and unpredictable in its details as are the capitalist processes of product innovation, technological and locational change'. Technology offers capitalists a vital means of control over the urban labour market and thus struggles over new technologies are crucial (Harvey 1985b). The coherence is inherently unstable. This is caused by the continual process of seeking out excess profit on the part of the individual capitalist, which pushes in two separate directions. The first is to search for it through the development of new technology or locations; the second is to 'convert temporary into permanent monopoly advantage'. It is the latter which becomes particularly important in the context of the urban class alliance activity within the urban labour market. This strategy involves capitalists in 'managing their own externality effects...to capture the benefits of the urban synergism' (Harvey 1985b). The difficulty with monopoly power is that it tends to slow down the dynamism of the accumulation process.

> 'Herein lies the basis of stagnation, then crisis formation. A sudden break with past technological mixes and spatial configurations often entails massive devaluations of pre-existing capitals and the breaking of powerful monopoly privileges' (Harvey 1985b).

So although stagnation is broken by this creative destruction it means a revolution in production process, the breaking down of monopoly privileges and new divisions of labour.

The urban class alliance will in this context, act in the interest of the dominant configuration in the locality. Because under Keynesianism, the local state has been crucial in structuring the dimensions of 'class alliances', they have been opened up to all sorts of infiltration from interests inimical to the further development of accumulation strategies. Because the alliances are so crucial to the demeanour of the locality and to a changing regime of accumulation, is precisely why the hegemonic project of Thatcherism has crystallised in the locality in the form of the UDC. Trying to fend off destabilisation, press for innovation and maintain coherence requires some control and co-ordinating force. It is here we encounter the

> 'art of politics...The politician of craft and genius can forge a relatively permanent and powerful coalition of interests so as to unify and articulate a sense of place bound community' and 'it is into this breach that the whole class of urban managers can insert themselves as a seemingly independent source of social power' (Harvey 1985a).

But how then can I explain alliances or conflicts across classes? Alliances in one urban region may consolidate in order to fight an alliance in another region. In the short term therefore, or within one region, this may seem like class compromise or the end of class based struggle in favour of functionally based struggles over the mode of production, but broadened onto the world stage and incorporating a periodisation using regimes of accumulation, it is possible to achieve a class perspective. The industrial working class operating in a region where there is a residual expanse of 'extensive' production, may cooperate with the locally rooted capitalists in order to gain advantage or hold their position with other coalitions involved in 'intensive' or even 'post Fordest' or industries elsewhere. This suggests that class struggle has not disappeared, but is taking place across multiple dimensions (Fig. 2.1).

In attempting to define class I argue that freezing it in one dimension is no use at all because class consciousness is comprised of residual elements from previous modes of production and class formation. This echoes Marx's conceptualisation of class contained within the Eighteenth Brumarie of Napoleon Bonaparte. I have argued that we must perceive class within a three fold categorisation: class as a historical phenomenon, in other words class resting on residual mode of production contained within a given urban region; class as a spatial dimension of the social formation, that is a categorisation of class which acknowledges the impact of uneven economic and social development and finally class as a fractional element of functional capital, that is giving primacy to production or circulation interests. This puts the state officials in contradictory locations, which are spatially specific and subject to the dominant group in the urban class alliance. This take us back to Harvey's notion of class alliance and structured coherence shaping and being shaped by class antagonisms.

The State in a Capitalist Society

All of this suggests that it is possible for the state to have become a crucial arena in which class conflict is not only played out, but becomes in itself an instrument of struggle. The insinuation of the expansionist Keynesian Welfare state deep into the social formation of every urban region during the post war era institutionalises the dimensions of the greater struggle which was already taking place.

But what exactly is the state? One answer to this question is that it depends from where you are viewing the phenomenon, at what point in time and from which perspective. Painter (1995) has recently explored some of the meanings of the concept of a state drawing upon Ratzel (the state as an expression of the joint social cultural and territorial interests of the people) to Hartshorne (the state as a territorial entity) to Marxists (the state as an instrument of class power or an arena of class struggle). More broadly pluralist and neo pluralist interpretations see the state as an arena where society's constituent groups, some more powerful than others, play out the drama of democracy on a crowded, but reasonably just stage (Dahl 1961). Neo-pluralist accounts of the state now converge with elitist and corporatists in assigning levels of influence and power operating through certain groups within the state (Jorden 1992). Corporatists much like Neo-elitists view the state as an instrument which is captured by dominant interests, coalescing elite labour and capital interests with state governments (Lindblom 1977). Again this view is increasingly discredited as corporate interests are increasingly marginalised from the state domain. The New Right is ever wary of the growth of a public bureaucracy and self-maximising bureaucrats (Olsen 1985).

For my own purposes I want to concentrate on the state as a moveable object. I will therefore conceptualize it as a series of 'institutions, apparatuses and processes' which constitute local and national governance operating within an international institutional framework and bounded by the democratic mandate of the British people (Duncan and Goodwin 1988). By separating these constituents I emphasise the different elements of the state form and suggest that it is important to examine how exactly they are are bound together as it is these bonds that are crucial determinants of power in any state form.

I suggest here that it is possible for the repertoire of state form to evolve alongside the changes in social formation aligned with the key periodizations in capitalist development (Aglietta 1978). Jessop earlier arrived at a similar conclusion:

> 'a distinct ensemble of institutions and organisations whose socially accepted function is to define, enforce and collectively bind decisions of the members of its society in the name of common interests and general will' (Jessop 1990)

Building on Aglietta and upon Jessop I want to explore two important issues in state theory: The level of autonomy from the economic sphere which the state enjoys and the importance of the separation of state powers in late modernity.

State Autonomy

The debate on state autonomy has raged since the 1960's and the protagonists within Marxism are exemplified by the instrumentalist Miliband, and the structuralist Poulantzas. The instrumentalist account views the state simply as a homogeneous, unified block of power which works on behalf of the ruling capitalist class: 'the executive of the modern state...is but a committee for managing the affairs of the whole bourgeoisie' (Marx and Engles 1975). Miliband saw the state as simply managing the affairs of capital, and of legitimating this rule with a panoply of legal, judicial and political powers

(Miliband 1968). In the modern state this was facilitated through the internal recruitment patterns which promoted the ruling class to prominent state positions.

This account ignores the way in which the state has been implicated in struggle against capital and the way in which the welfare state developed in the post war era.

> 'The state function is simply to co-ordinate and manage the economic crisis generated by the mode of production through political and ideological interventions' (Miliband 1968).

Poulantzas in contrast argued that the nature of the personnel in the state apparatus is less important than the stance which the state takes in periods of crisis or when faced with intra-class conflicts between different fractions of capital. He argued that agents are simply the bearers of social relations by stating that:

> 'at all times premises of action exist which are not at the disposal of actors themselves; action is tied to inputs which cannot be modified or expanded' (Poulantzas 1975).

He suggests a dynamic state which balances class support across fractions, and evolves policies which adapt to deal with 'problematic movements'. He emphasises its relative autonomy which is essential in allowing the flexibility which is needed to balance out the contradictions of capital accumulation. He affirms the importance of the appearance of state neutrality. Most importantly, writers like Poulantzas see the nature of change in state institutions as a mirror of class struggle, perceiving institutions as being constantly captured and recaptured by opposing class forces. Poulantzas suggests that it is not the capitalist class that the state exists to protect, but rather the accumulation process (Poulantzas 1978).

Following Poulantzas I believe that the state is not allied to a class so much as allied to the accumulation process. Indeed the modern state is, itself dependent on this process (Harvey 1985 a). It seems to me that at crucial points the state is more or less autonomous, in relation to the capitalist class, whilst always having at its base the logic of capital accumulation. This again turns us towards class fractionalisation which in the context of relative autonomies suggests that the state becomes a focus of conflicting interests.

But how is this logic translated? What determines the level of autonomy? Most importantly how is this autonomy translated into the range of relationships in which the state is engaged? How does one part of the state relate to another? How does one instrument of the state relate to another?

Central Local Relations: The Implications of the Separation of State Powers in Late Modernity

Much of the recent discussion around the spatial dimensions of the state has been conceptualised in terms of central local relations. The conventional view on central local relations until the 1970s, saw local authorities either as a partner of or as an agent for central government policy. The former conceived of the two elements of the state under the power of Parliament while the latter saw local government implementing policy on behalf of central government (Dearlove 1979). The simplicity of this formula belies a more complex reality. And the uneven development of public policies, as well as that of regions, has now undermined the pluralist public administration theorists who promoted this idea.

As the effects of economic restructuring began to be felt in the UK's inner cities, the solutions to urban problems offered by central government often became the means of politicising and channelling the energies of urban populations (CDP 1975). Campaigns were often directed against local authorities over issues of collective consumption, such as housing. This was possible because one of the effects of the welfare state had been to widen the issues which were seen as legitimate arenas for struggle, and as Manuel Castells argued, issues of struggle over collective consumption in its turn, widened and deepened cleavages within the working class, extending the notion of class struggle (Castells 1977).

Local authorities were a focal point of struggle, partly because they delivered services and were peoples' main point of contact with the state and partly because they were in the curious position of being an instrument of economic and social discipline on behalf of the central state and a focus of democratic articulation for local people (Duncan and Goodwin 1988).

In the late 1970's Cynthia Cockburn saw local Government as an agent for the capitalist state. Her study of Lambeth council has been criticised, for its instrumentalism and for ignoring struggle within the state arena (Cockburn 1977). Had it been written just four years later, the conclusions might have altered radically. In this sense her study was quickly outdated and her critics had a point. Yet her attempt to focus local/central relations within a Marxist framework was useful. The implication that the local state had a part to play in capitalist accumulation may have been simplistic, but it was by no means totally incorrect. My difference with her thesis centres on the simple 'transmission' notion from central to local and the institutional confines which binds her theory to functionalism.

The local and central state had different functions, tasks and roles, and later theorists using a corporatist framework suggested that a dual state was a more appropriate explanation for the roles of local and central government. Cawson and Saunders felt that there was a division between the activities and responsibilities of the local and central state (Cawson and Saunders 1983). This saw a central state which concentrated on production, where decisions were made in a corporatist environment, linking economic decision-makers, industrial and finance capital, the white-collar public sector and the organised trade union movement in alliance with central government. The local level was involved with consumption, where there was a competitive political struggle over consumption issues.

This obviously reflected some of the social cleavages which Castells mentions. In some sense this debate was derived from the work of O'Connor (O'Connor 1973) in which he had specified functions for the different elements of the state, social investment was related to production capacity in fixed capital and social consumption was labelled collective consumption or consumption fund. His suggestion was that there is a perpetual tension between these two, and that the local state is the arena in which the struggle around collective consumption takes place. There has been a lot of criticism of the dual politics model, the main thrust of which is that there is far more overlap than the model generally credits in production and consumption capacity of either the local or the central state. Again this criticism became even more pronounced as the eighties have progressed.

A Marxist framework and a spatial analysis were united in the Duncan and Goodwin study of central/local relations. Duncan and Goodwin distinguish between 'state apparatus' which they see as state institutions and 'state' which they perceive as being comprised of the social relations. They suggest that relations are unevenly developed, hence on the one hand there is a need for different policies in different places, and on the other a need for local state institutions to formulate and implement the centre's policies (Duncan and Goodwin 1988). Although they incorporate spatial analysis into local social relations and acknowledge Harvey and the notion of structured coherence, they seem reluctant to jettison an instrumentalist approach, instead seeking to 'bolt on' space. They differentiate between causal and contingent local processes. However they emphasise the dialectical element of the relationship by saying that the local state is both the agent and obstacle of change and that the local state provides both a representative and an interpretive role. They identify truly successful local struggle as when an oppositional interpretive role combines with a widespread representational role. The centre then has to move to neutralise this and local autonomy is removed. Their contribution is to link uneven development to social change, and link the state to this process by saying that it plays a part in uneven development by institutionalising coherence.

The Duncan and Goodwin study was a response to two imperatives: the need to explain the drastic deterioration in the relationships between local governments and the central state which took place through the 1980s and the simultaneous pluralisation of local models of local government good practice. Stoker sustains this argument in his contribution to the debate by showing the increasing levels of central intervention in the local democratic arena during the 1980s. He suggests that this occurred through the use of five strategies of intervention: legislation, minimal consultation, targeted funding, the by-passing of local government and a major reorganisation of local government (Stoker 1988). Despite this acknowledgement of the centre's impact on the periphery he has also emphasised the importance of 'moving beyond a narrow focus on relations between local and central government' (Stoker 1988). His questioning of the idea that we can simply deduce a separation and compartmentalisation of both the institutions and the process of the state on the basis of a spatial differentiation is a useful bid to take us beyond these polarities.

The state in late modernity is complex and it is infused with manifold decisions about capital accumulation strategies operating across different circuits of capital. This means that different tiers of the state are held up with different moments in the circulation of capital, and that even within tiers there are now massive tensions between different interests and fractions. Therefore a 'dual' state or local/central relations is not sufficiently detailed or nuanced to deal with the reality.

A Theory of the Contemporary State?

Much of the recent work on the British state using a regulation approach has been done by Jessop (Jessop 1990). He shares a commitment to conceptualizing economic, political, ideological and social relations by using the coupling of 'regimes of accumulation' and 'modes of regulation' and like other Regulationists, uses this coupling in order to identify the nuances of relations across and within these arenas. In doing this Regulationists have shifted the ground substantially from the simplistic base/superstructure analysis in which the earlier debates on the state were trapped. Regulation theory attempts to deal with the fundamental question of how capitalism survives so many of its own crises and continuous contradictions and reproduces its ability to continue the process of accumulation. Aglietta was the first to propose that it can be explained by using the concepts of a 'regime of accumulation', that is a combination of production and consumption activities which fuels the accumulation process and a corresponding and complementary 'mode of regulation' which comprises the institutional and societal patterns of organisation which deal with the problematic of class struggle (Aglietta 1979). The contours of this struggle are defined by the relations of production, circulation and exchange. The impetus in the struggle is to find complementary regulatory mechanisms and strategies to fuel

Figure 2.4 Regimes of Accumulation and Modes of Regulation

Regimes of Accumulation	Modes of Regulation
Extensive Accumulation	Old Regime Regulation
Intensive Accumulation – Taylorist (without mass consumption)	Competitive Regulation
Intensive Accumulation – Fordist (with mass consumption)	Monopolistic Regulation
Emergent Post Fordism (not yet developed)	Emergent semi flexible Mode of Regulation

accumulation. The combination of a 'regime of accumulation' and 'a mode of regulation' are pivotal to the developmental periods of the overall capital mode of accumulation.

British regulationists are interested in how the institutional and processual form of the state and civil society changes in conjunction with massive transformations in the economic sphere. Therefore, put simply, the question is how a dominant paradigm comprising rules, norms, institutions and social relations gives way to a new paradigm which is recreated through struggle arising from a crisis tendency. Shifts can take place within the dominant paradigm and from one paradigm to another.

The shift which most authors have focussed upon is the move from a fordist, (Keynesian) paradigm to a post/ neo fordist (neo-Keynesian) paradigm (Amin 1994). Fordism has been conceptualised as the development of mass consumption markets which serviced the productive capacity engendered by mass produced assembly techniques based on Taylorist principles. In practice this meant intensive mechanisation of long production runs, separation between conception and execution of development of products, the use of semi skilled labour, with the increasing fragmentation of tasks. The productivity gains and reduction in real costs allowed wages to rise without harming profit. International fictitious capital allowed large corporations to seek geographically dispersed markets. Fordism is associated with US hegemony which reached its zenith in the post war reconstruction process.

There is some debate as to how realistic it is to refer to Fordism when capitalism appears to accommodate a multiplicity of modes of production (Amin 1994). How extensive do patterns of Fordest industrial organisation become within the international production process, and what exactly constitutes the period known as Fordism? One of the key problems for urban analysts is to disaggregate the Fordist production period from the Keynesian state management period in order to conceptualize the relationship between the two spheres.

Nevertheless many British neo Marxists have been unequivocal about the nature of the Fordist project (Hall and Jacques 1988). They categorise its hegemony as beginning during the second decade of the 20th century and lasting until the 1973 oil crisis. They do not differentiate between Taylorist production methods within factories and the widespread facilitation of Taylorist methods through state Keynesian demand management techniques as the Regulationists do. According to their analysis, the current period is post Fordist. They use the Regulation approach to explain new life styles and consumption strategies as well as changes in social relations.

Aglietta, however, suggests that Fordism began in the first decades of the 20th century but stresses that the crisis which took place in 1973, which developed as an accumulation crisis in the capitalist economy, and a fiscal crisis within the capitalist state, signalled a move towards neo Fordism (Aglietta 1978) In other words this transition is not perceived by Aglietta as a post Fordist movement – a new regime of accumulation – but as means by which intensive accumulation could continue to take place.

Jessop concedes many of the contradictions within the regulation debate and attempts to deal with it in part by specifying what the minimum conception of a Fordist period would require in terms of structural and strategic elements (Jessop 1994). He has suggested that there are five factors or units of analysis which can be used to define the overarching paradigm. The first three are: the Labour/production process, the Accumulation Regime, which he describes as the macro economic regime sustaining growth in capitalist production and consumption, a Social Mode of Regulation, which he characterises as an ensemble of norms, institutions and organisational forms, social networks and patterns of conduct which guide the regulatory regime. His fourth unit is a mode of Societalisation, which is a pattern of institutional integration and social cohesion which complements the dominant accumulation regimes. The final unit is the Social formation which he suggests comprises all of the preceding elements.

Like Painter, (Painter 1991) Jessop is more interested in the mode of regulation than he is in the specificities of the regime of accumulation and it is using the mode of regulation to investigate the state that has recently preoccupied both theorists.

A mode of regulation has been defined as

> 'an ensemble of norms, institutions, organisational forms, social networks

> and patterns of conduct which will sustain and guide accumulation regimes' (Lipietz 1986).

This, of course, includes the state apparatus. The mode of regulation is important in ensuring consistency with the regime of accumulation in order to ensure that crises within the mode of development perform a regulating function only. Theorists vary on the primary force or object of regulation. As Jessop stresses, for some it is the production process, for others, it is the wage relation, and for others again it is the social formation (Jessop 1992). The transition in the field of production has been discussed in terms of the move from Fordist to post Fordist systems of production. Within the state form it has been termed a move from the Keynesian to the post Keynesian state.

A Framework for the Operation of the State?

Is there any way then that we can combine Harvey and the Regulation School to give us a spatio temperal foundation for the analysis of state politics in the city today? Any such framework would probably look something like this:

The industrial city was a crucible for capital accumulation, transforming the nature and role of the urban class alliance and with it the local state and political formation. The capacity of the 'class alliance' to wield monopoly control inevitably diminished as global capital movements and international credit markets transcended spatial and temporal boundaries. Both labour and capital entered a new phase developing new roles within this new industrialism as it has done at each major juncture in capitalist development.

The industrialised city underwent successive crises of overaccumulation, massive immigration and emigration of labour, and tremendous advances in technological change on its journey towards a fordist/keynesian configuration. The crises of overaccumulation were tackled by the option of both a 'spatial fix' and the pouring of surpluses into the further production of the built environment, in the form of fixed capital as well as the consumption fund.

Harvey's conceptualization of the relationship between the productive and economic sphere of Fordism and the social political framework of the Keynsian state differs from the Regulation school. Harvey argues that Fordism failed and had to be converted into state managed Keynesianism. The flip side of overaccumulation was underconsumption, which stalked the United States in particular during the 1920-30s. The trick employed to solve the problem was to move from supply side to demand side urbanisation, with the help of finance capital. International capital facilitated debt financed infrastructures and easier sectoral and geographical switches, but tied the urban environment more closely to capital flows, making urban areas particularly vulnerable. The advantage was that: 'the proper allocation of credit to production and consumption held out the prospect of balancing both within the constraints of continuous profit realisation' (Harvey 1975a). Of course there were problems with the new international credit system. Financial capital automatically centralised and consolidated a geographical hierarchy and it had to be regulated otherwise wild speculation would create crisis.

Harvey argues that this is precisely what happened to spark the Keynesian strategies of fiscal and monetary demand management, which in itself consolidated demand side urbanisation. The Regulation school argues more generally that Keynesian demand strategies were the stimulus to and solution for Fordist production methods and that finance capital was a mechanism with which to control production and consumption flows. Under either formulation the role of the class alliance in the post war period has accordingly shifted focus, moving from a concentration on production interests, to a more varied and loose coalition of interests around consumption and distribution issues.

This is key to anticipating the role and the effectiveness of the operation of local government, within the industrial and increasingly post industrial state. Local government in the Keynesian welfare state has become critical to the demeanour of the urban class alliance. Its strategies and strengths, its ability to negotiate and bargain across fractions, is crucial to transforming accumulation strategies and regulatory regimes.

By the 1960s the city as a centre of consumption and the realisation of profit, vied with the city as centre for production and the realisation of profit. The effect of this was to create new sources of tension and new class cleavages which the class alliance moulded and sought to manage.

This was exacerbated when crisis reverberated through the western capitalist system in 1973. Interpretations of the crisis vary, as do interpretations of its effect upon the dimensions of the state form. Firstly, Aglietta dissents from others in his view that the crisis did not necessarily mean a paradigmatic shift in the regime of accumulation. Secondly, there is still significant debate as to whether the post '73 period can be categorised as neo or post Fordist (Amin 1994). Finally, there is even more controversy about what neo or post Fordism actually means in terms of a regime of accumulation and a mode of regulation (Peck and Tickell 1994). While there is a measure of agreement about the nature of the crisis therefore, there is as yet no common line on what comes after it.

Harvey suggests that temporal displacement led to extreme indebtedness and inefficiency pressure, spatial displacement of Fordism was ever forced to more intensive levels of devaluation and destruction, and highlighted a concern with the circulation of revenues. This had an effect on the role which local governments played in their localities as they sought to maintain their hold on central revenues.

The redistribution and consumption strategies of the Keynesian state did not sit easily with the production strategies. The production of urban infrastructures was underpinned by powerful interest and growth coalitions which managed this particular type of urbanisation. The locality had substantial interest in two features of public expenditure:

1. Redistribution of social expenditures.
2. Public expenditure in the built environment.

Both of these had to be maintained. The question was how to do this – and who could do it?

The fiscal crisis of the state (O'Connor 1973), was underpinned by the overaccumulation of assets in the built environment and social expenditures, showing extremely low rates of return. The problem for urban class alliances, which were previously managerialist, was that they had to convert into something different. This was difficult as they were so deeply embedded in the state monopoly tendency. But stagflation could be resolved only through a closer equilibrium between the production of surplus for their real as opposed to fictitious value.

It seems plausible to argue that the state was a centre of the mode of regulation within the project of Fordism. It had a significant task in regulating demand in order that consumption could satisfy the production imperative. The state managed aggregate demand so that firms could plan ahead and invest in Research and Development over long planning periods. The state was crucial to the operation of organising consumption strategies and providing the framework for regional and planning policies as well as physical and collective consumption infrastructure throughout the entire period.

Even so, as a result of mobility of international capital and the product of labour and capital surpluses, international competitive strategies had preoccupied policy-makers in the early seventies. Subsidies, the preparation of land and property and the state or the regional state acting as a collective entrepreneur were present in the consumption enclaves of the western capitalist world. In the current period (from 1973) this has intensified and sharpened and Harvey is eager to extend his analysis of the political and economic machinations of urban capital within this period. Here I bring his analysis of entrepreneurialism up to date. Thus fiscal contraction and strategies to deal with the crisis in profitability, the changing nature and function of the state in relation to the accumulation process and a switch in the function of the urban region in terms of capital displacement and a search for innovation are the key features of the state economy relations in the present period.

3 The state and policymaking

So far I have documented the evolution of the UDCs as a policy instrument of urban regeneration and I have contextualised their implementation against the backdrop of political and managerial relationships between different levels of the state. In the previous chapter I sought to provide a functional and latterly, a sequential framework in which the UDC could be interpreted on a theoretical plane. The problem with the framework so far is that it lacks the ground level empirical conceptualisation that I ultimately seek to provide. This is why in this chapter I intend to move into the field of policy analysis and it is where I will put in place the final tools with which I intend to examine my subject.

Which policy models might be useful in interpreting how the dimensions of the policy process at the micro level might reflect the nature of institutional and organization development at a meso level in the state and the structure of economic forces at a macro level more generally in society? If there are models that are useful in interpreting these relations, what should they be telling us about the ground rules that are shaping their design? What do they tell us about the forces that drive them and about the obstacles or contradiction that they encounter? Finally what contemporary models of process are useful in sketching governmental relations in today's state?

Harvey and the Regulation School have coalesced in recent work in their attempt to periodise the transformation in economic and social paradigms and thus state activity (Harvey 1989a, Offe 1985). Recent policy scientists such as Pollit, have echoed this theme and have explored a variant of these periodisations in the development of public policy initiatives in the Keynesian Welfare state (Pollit 1990). This approach is a useful addition to the toolkit of policy analysis and such a conceptualization of periodic change straddling the economic and the politico-administrative sphere in tandem, assists us to speculate on the linkages between process forms at the level of implementation and middle and macro level environments. Such a periodisation of policy styles and frameworks, showing the dominant, though not all encompassing, trends in institutional policymaking, is helpful in sketching the configurations of state/economy ensembles in urban regions. It also serves as a useful way of interpreting the prescriptive and descriptive capacities of policy in the post war era in particular.

Keynesian Rationalism

For most of the post war era the rational model of public policymaking has been the normative optimum from which bad policy deviated (Simon 1947). It was contested first with an incremental model (Lindblom 1959) and later with a range of models combining rational and incremental approaches (Dror 1968). The rational model established a complete set of operational goals which were then assigned priority after an inventory of values, resources and a comprehensive set of alternative policies was prepared. Each would then be comprehensively cost benefited, net expectations compared and alternatives established (Rhodes 1981). The model is an ideal type in the Weberian sense, but it nevertheless became the benchline for the measurement of all normative processes of state policy-making. It had an important legitimating function in the arena of public policy-making between the 1940s and the 1960s and was temporally linked to the expansion of the state in both the United States and Europe.

The rational model could only be seen as an optimum during the era when 'Fordism' was at its height, complementing and reiterating Keynesian state practices. The mirage of stability upon which post war policy planning depended, incorporated rationality not only as a goal, but as a real possibility. Crisis was not part of the organisational vocabulary. The apparatus of the state was viewed as controllable and manageable. The rational model therefore served as an ideological tool, it depoliticised the decision-making environment, making it a techno-rational environment, whereby all policy would be judged by the criteria as given. It obscured values, it obscured power relationships and gave a chimera of legitimacy and widespread accessibility to the public policy process. By the late sixties this was no longer a sustainable image within the policy field.

In the search to find the answer to the problems then experienced in policy making, which were, I would argue, system faults, two themes emerged. What and who drives the policymaking process? Here the focus was on the relationships between the state and the economy (Marxists), between various interest groups (Pluralists and Neo Pluralists), between powerful subgroupings within specific strata of society (Elitists and Corporatists) and finally between groups of self maximising bureaucrats or agents (Neo Liberals). Secondly the question was how do we discover who/what drives the process? Here there have been three approaches, the top down rational /managerial paradigm (Hood 1984) and the bottom up strategic mapping or implementation paradigm (Barrett and Fudge 1981) and the power dependency network approach (Marsh and Rhodes 1992). The rational managerial paradigm relegated politics to a subsidiary position within decision-making. It did not take account of the way that political goals were turned into action in the process either overtly or covertly. The critique of this approach was a particular feature of the non decision-making literature (Bachrach and Baratz 1963). Neither did it account for bargaining and negotiation processes in the way that contemporary studies do (Hanf and Toonen 1987). The later studies began to penetrate the role of interests in the policymaking process, but rationalists saw such work as deviating from the norm of rational policymaking rather then seeing it as being integral to the process.

Incrementalism was perhaps a better descriptive tool, but it hardly served any better as an explanatory device for the problematic of policy design. In the 1950's, Lindblom's work attempted a number of things: to make explicit the part that values play in the process, a recognition that when policies appear to be failing, small 'incremental steps' can be taken to 'adjust', that policy is 'serial' and that it is made with reference to and by a number of agencies. Rational and incremental formulations were developed by policy scientists who were intent on improving public policymaking in an era of apparent consensus. But they assumed that policy could be made more rational, simply by improving its design.

Following Offe I would suggest that the reason why nobody could get policy right and indeed why policy scientists could not find an adequate explanation for the difficulties inherent in the policy process is because of the contradictions which the accumulation process places on the policy-making arena.

Offe argues that the capitalist 'system' produces inevitable obsolescence within its administration when 'normative schema' accompanying a specific accumulation regime become outmoded and have to be replaced with a series of 'strategic schema'. The 'dysfunction' of the system, is actually a result of the contradiction between capital accumulation interests and state administrative interests (Offe 1984). There is always and indeed necessarily, a lag between the activities of economy and administration. The administration faces two imperatives, legitimising its own actions through the representation of 'common interests,' and protecting the accumulation process in the interest of capital. However impossible it is to deal with the contradictions, the state usually attempts to impose some remedial action to bridge the contradiction, often by reordering the institutional ensemble. Therefore, the rules of organisational change are felt, by Offe, to be a result of the necessary dysfunction between the administrative rationale of the state and the accumulation rationale of capitalism. Decision rules are therefore crucial to observing the way in which power is transformed in the process and whether we are undergoing, in Jessop's terms a strategic or a structural change (Jessop 1992). What happens when the 'rules of the game' change, for Offe, is that the innovation within state strategic dimensions becomes the innovation of those dimensions. Offe suggests that the rules in this situation are generated by the state itself and that the institutionalisation of these public policy rules poses a key problem in public policy formation.

The question is, just how does the state contribute to the maintenance of capital accumulation and what different strategies are necessary in different circumstances? The contribution which the state makes to the accumulation process, is in late modernity, so varied and so integral, that it would take another book to document it (Sadler 1992) (1). In this section I want to concentrate on the strategic dimensions of this relationship. Offe suggests that there are three functional imperatives in the relationship between the state and the accumulation process: exclusion, maintenance and dependency. The crucial task for the state, is to maintain co-existence and harmony between these functions. The force that is necessary to do this is in a liberal democratic state is legitimation. Thus we are given a fundamental problem for the state which shows itself clearly in both the policy development and the mode of decision making of the state.

Legitimation is crucial because if the capitalist liberal democratic state is to function as such, it must conceal its true nature (Offe 1985).Thus the question of legitimation in the state form and its relationship to policy becomes central to examining the way in which policies are determined and implemented. This suggests that the accumulation is a powerful constraint on the policy process and that it depends on the balance of exclusion, maintenance, dependency and legitimation. (Offe 1985).

The conflict between the state as an efficient vehicle for capital accumulation and the state as a legitimate form of social administration is most acute in times of crisis as we can observe from the literature documenting the 1970s crisis. At these times the dual role of the state as conflict generating (its structure) and conflict resolving (its capacity) becomes impossible to negotiate (Offe 1985). The contradiction that exists between the structure and the capacity of the state suggests that every time there is a difficult issue for it to tackle in terms of problem articulation, there is also a problem to deal with in terms of its mode of operation.

The difficulty faced by the state and by successive public policy paradigms is to maintain the 'balance between internal rules and functions required by the accumulation process'. (Offe 1985). When crisis occurs in this dimension Offe suggests that the institutional functions which channel conflict articulation and conflict resolution are usually reduced. The balance is unstable in the present period because accumulation require competitive strategies, minimal state expenditure on social strategies and state intervention. But it can also require new and innovative policy ideas. The question is how to do it legitimately. If the policy process becomes unmanageable this would presage a change in the policy-making structures. The administrative strategy then turns to the development of new channels of conflict articulation that develop and absorb the energies of the people. If the new channels fail, governments rely on the limitation of the democratic political process. Offe speculates that this situation pushes the location of political conflict and the institutional location at which state policies are formed away from the traditionally accepted points of operation (Offe 1985).

Later, he speculates on the capacity of an administrative system that could cope with dual demands. The way to manage this he suggests, is for the state to centralise and decentralise at the same time: 'the state must increasingly organise and regulate the socio economic functional coherence of the whole order and on the other hand lead a distinct and limited existence' (Offe 1985). The neutrality of the state as a 'substitute consensus' is one way of dealing with the difficulties of reconciling this conflict and this would appear to be the dominant characteristic during the height of the Keynesian Welfare state. It is the extensive use of 'para parliamentary modes of decision-making' which marks out the present period. This rests on the use of consultancies, specialists and people with particular knowledge. Offe implies that elites are constantly searching for non-political forms of decision-making and that citizens in response do the same, resulting in urban social movements. Parallel retreats from official channels of policy-making occur:

> 'as a consequence of this, struggles over substantive issues and politics, the institutional form of conflict resolution degenerates into informal and mutually disconnected modes of struggle and decision'
>
> 'decision-making power depends on the presence of continuity in the process – thus the criterion of stability of accumulation is incorporated into the pursuit of interests and policy' (Offe 1985).

The state does then become a source of power in itself:

> 'a specific set of institutional forms which are institutionalised in the state and which generally define to what extent and through which channels citizens as members of classes and groups are allowed to participate in the formation of public policy' (Offe 1985).

Offe therefore takes a similar position to that which Poulantzas developed earlier, the state necessarily separates itself from the economic processes of production and reproduction to legitimise its practices. Offe's conclusion about the way in which the state chooses to do this goes some way to explaining the variously held ideas about growth coalitions but tiptoes also towards Harvey's conceptualization of an urban class alliance:

> 'My thesis is that today's public administration is faced with a situation where many of its operations in which the competition of state planning function can no longer be accomplished by itself but must involve individual citizens and their societal organisations as sources of assistance in the authoritative performance of the executive function' (Harvey 1985b).

Offe argues that the design of decision rules is enmeshed in the divergent rationalities of the administration and the accumulation process. The eventual form which the decision rules take, in the manifestation of policy frameworks and styles is not something that is static, it is dynamic. Rules are not shaped as a simple transmission of the needs which accumulation processes project, but rather are shaped through power struggles for which they are a channel. The distribution of rules and their effect on the distribution of resources is a function of social action, the interaction of actors and the (state) organisation.

In another essay Offe suggests that there are different degrees of interest group distortion in capitalist society (Offe 1985). The distortions he suggests are not randomly distributed but are structurally contingent upon class position. The differences in the organisational practices, problems, strategies and structures that we observe between workers and capitalist interest groups represent asymmetrical interests. That is, the rules and resources implicit in interest group activity are weighted by class position and are therefore unequal.

The political forms which emerge within state processes are therefore schemes of recognition of that conflict, and conflict occurs simultaneously on two levels: class conflict within and class conflict about political form. Offe suggests that the first takes place between groupings that are able to participate in and who have 'a knowledge of the rules of the game'. The second, he says, is hidden under the mask of neutrality of process. It is about the hidden way in which process forms are to be attacked or defended. This means that process forms are inherently unstable within liberal democracies. This is difficult for both practitioners and theoreticians. Perhaps this is also an indication why a plurality of approaches have been taken towards process form and why the rational models are distinctively unyielding as a framework of analysis. It means that the state is constantly searching for a way in which 'continuity and change' as well as internal 'contradiction' can be solved.

Mapping conflict within the state form is difficult enough. Disaggregating it sufficiently to separate conflict about form from within form is even harder. But it is only by doing this that we can observe some or no change of paradigm.

Policy analysts started to look at this question in the late seventies and early eighties, particularly in North America. This view of the policymaking processes is even more important in contemporary Britain. It developed from a new approach to implementation which focussed on the bottom up strategies of actors and agencies in order to explicate the 'implementation gap' (Pressman and Wildavsky 1973). This is most helpful in the context of analysing the UDC because it allows our own focus on micro processes to marry with our complementary top down or external to internal route into an examination of the forces that shape UDC policy.

The pluralist interpretation of power relations in a multi agency environment is articulated as a process of 'partisan mutual adjustment' which seeks multi participant consensus. Lindblom's theory of 'mutual adjustment' came from his analysis of the co-ordination of agencies or people in the absence of central co-ordination.

> 'One can endorse partisan mutual adjustment indiscriminately because of its merits in dispersing otherwise dangerously concentrated power' (Lindblom 1965).

But ultimately 'partisan mutual adjustment' may explain the consensual elements of inter-agency politics in the post war period, but it does not explain the conflictual situations, which existed then, but are more prevalent now. Ultimately empirical work has overtaken the model, for example, Rhodes investigation of subcentral government (Rhodes 1988).

While Marxists during the 1970s and early 1980s concentrated on the larger structural questions, much of the detailed work on policy making was still being done within a pluralist framework. Neo-pluralists wrote of a national local government system or a national system of policy networks (Rhodes 1986 and 1988).

Rhodes in particular developed the notion of a national local government system (1986), and later a more complex conceptualization of central local relations which attempted to analyse the state using inter-organisational theory (Rhodes 1988). Here he looked at the distribution of rules, power interests and values in agencies and the way in which these factors interacted with organisational goals and resources, interorganisational relationships and the level of discretion in organisations. In this he converged with the implementation theorists (Benson 1983). Rhodes' explication of policy networks unveiled the complex layering of interests within the state form. Indeed his forays into the world of producer and issue networks are illuminating in uncovering the interconnections which exist, as well as the contradictions between the multitude of interests that are involved here. Comprehensive as his study is, it remains entirely within the institutional circumference of the state. Here I want to draw out his analysis and extend more deeply into the interagency networks which surround the state.

The networks which Rhodes describes give a far better vantage point to view the power relations between the different elements of the state but they are still without a necessary contextual framework. As the boundaries become increasingly blurred between the private and public sector contributions to state policymaking, this becomes a problematic issue in examining contemporary state processes. The task 'apres' Rhodes is to give cogniscance to the interface between these networks and the locality in order to provide structural explanations for their actions.

Barrett and Fudge were preoccupied with implementation in part for similar reasons, suggesting that difficulties of process trajectories derive from gaps perceived between objectives that are set at a particular point in the policy process and outputs and outcome that are created at another point. Although they rightly eschew a chronological ordering of the policy process, conceptualizing it instead as a set of interactive and interlinking procedures, their typology of elements seems to act upon the process rather than seeing the process as a force reflecting the relations of production and reproduction and participating as a channel through which those forces are shaped (Barrett and Fudge 1981).

Political economy, they felt, was too broad a perspective to be of any use in determining what exactly the policy process is about, but they obviously recognise that the investigations into micro structures and organisational systems cannot fully support a framework of any intellectual weight. Therefore they fill the conceptual gap by concentrating on bargaining, negotiated orders, linkages and the dividing line between consensus and conflict. This takes them to the point that Wildavsky had reached when he describes the policy process as a

> 'confluence of partial understandings among interested participants that changes with unfolding events' (Pressman and Wildavsky 1973).

However, they do point us in the direction of the role of values and the power structure of the policy process. In seeking to determine how these factors invade policy-making as well as determining the structure of policy-making the most helpful definition of process has been provided by Healey:

> 'Policy processes are thus transformations through which political pressures, ideas and problems are converted into policies and policy

> measures, competencies for implementing such programmes delineated and operating practices developed' (Healey 1990).

Healey identifies the key factors which policy definitions have failed to negotiate: the place of ideology in the process, the identification of interests in the process form, the inherent tensions which ineluctably lead to an implementation gap and the part that structuration plays in the process. In fact, Healey's somewhat uncomfortable cohabitation with Offe can lead us in the direction of an empirical and conceptual framework for the analysis of policy, but before embarking on that task it behoves us to examine the multi levelled institutional and conceptual terrain in which policy processes are entwined.

Benson clearly offers us some conceptual tools developed in his analysis of interorganisational working (Benson 1982). He is interested in what co-ordinates and influences policy across agencies, what structures them and how resources operate, influence and interpenetrate, and how interests are built into a particular policy sector. Benson defines

> 'a policy sector as a cluster of complex organisations connected to each other by resource dependencies and distinguished from other clusters or complexes by breaks in the structure of resource dependencies'(Benson 1982).

He assumes that it is important to examine rule making across three levels. First the administrative structure, that is the surface level of linkages and networks held together by resource dependencies. Second the interest structure, that is the set of groups whose interests are built into the sectors. Third the rules of structure formation, i.e. the relationship between rules and the process of capital accumulation. Therefore Benson echoes Offe to some extent in perceiving the plurality of operational levels at which the state functions as a way of resolving functional and ideological contradictions.

Latterly the rules which Benson postulated have assumed greater importance in contemporary work on policy processes carried out by theorists influenced by the Regulation Schools. Those who have attempted to look at the state form have been interested in the first instance in unpacking the nature of economic state relations, taking us back to question of authority. Stoker has used Regulation theory to look primarily at the state at the local level, his work is one of the few applied attempts to analyse the state using a Regulation framework (Stoker 1989b). Even so he conceptualises the state/accumulation relationship in terms of a 'certain correspondence' in periods of stability and growth. Stoker takes this analysis a little further when he conceives of a role for the state during the period known as Fordism. He suggests that the role for the state is about linking patterns of consumption to mop up mass production activities and to consolidate a mass production and consumption market by facilitating social reproduction, i.e. health care, education (Stoker 1989b).

Jessop similarly draws a correspondence between the nature of the links between the Fordist labour process, regime of accumulation and the social mode of regulation and societalisation, and those units of analysis under post fordism. A fordist mode of regulation would include: state sponsored social reproduction, state mediation of conflicts between capital and labour, protection and underpinning of tripartite industrial relations structure, buffering of the cyclical movements in competitive markets and stability for investment strategies of capital. A Post Fordist system would underpin/encourage the conversion of productive and reproductive processes to flexible templates, increase the availability of new technological infrastructures, encourage an innovative pattern of accumulation, polarise the position and incomes of skilled workers and service classes and encourage interaction with global competitive markets (Jessop 1994),

Although it is perhaps the only option we have of observing the nature of the state/accumulation relationship after Fordism, can a post fordist state simply be adduced from its respective role under Fordism? It is much more difficult to predict the nature of the post Fordist state in the current organisational flux, particularly when there is clearly no new 'regime of accumulation'. Stoker begins to distinguish the role of the state in a transitional period. He highlights two functions which are important in the present period, the part that the state plays in facilitating economic and social restructuring and the management of the social costs of transition (Stoker 1989b)

By attempting to trace the dimensions of regulatory systems both Stoker and later Painter (Painter 1992a) pick up on aspects of uneven development and how this might affect regulatory systems. Stoker looks at some of the 'filters' which affect strategies on a cultural or historical axis. Painter concentrates on spatial characteristics, emphasising the likely importance of uneven development in shaping the post Fordist world. They both emphasise the contingency of regulation on the outcome of social struggle.

Hoggett has also chosen to analyse local government using Regulation theory from a different approach (Hoggett 1990). Rather than looking at the tasks of the state within a transitional framework he draws on organisational analysis to focus on two features of the 'deep structure' (2) of post Fordist forms of regulation, flexibility and the centralisation of command and control, and the decentralisation of implementation in organisations. He highlights the role of the new information technologies in facilitating the structures of regulation and he suggests that these structures cross previously private/public boundaries. He does however distinguish the precise effects of such strategies in the state sphere by looking at three public sector methods of 'allocating services' which he speculates are critical to the success of a new mode of regulation.

The first is the introduction of the price mechanism into public sector services.

I would argue that there is evidence to support this contention and that such mechanisms can be seen in cost benefit analysis, programming, budgeting, review systems and the more current economy, efficiency and effectiveness criteria. These can all be conceptualized as mechanisms which attempt to 'determine the appropriate basis for decision-making in the absence of clear and unequivocal profit signals'. Increasingly neo liberal approaches to the public sector seek to incorporate the equivalency of a price signal in relation to monitoring and evaluation strategies. Secondly Hoggett talks about the state's increasing reliance on technocratic processes managed by specialist communities. This is a process of which the UDCs are an example. Finally he discusses the restructuring of the democratic arena for debate, again an area of significant importance within the UDC.

What then are the Regulationists offering to the debate on the policy process of the state which makes it attractive within the framework that I am developing? Firstly they acknowledge there is a link between state and economic transformation and in doing so acknowledge the fluidity in state form. Additionally they seem to acknowledge the importance of a range of different processes operating at different levels and suggest that these might to some extent correspond to the social processes that are taking place around them. But the question is how these processes can be revealed. Jessop suggests that it can only really be done by a 'detailed analysis of the state forms and political practices'. Having said this Jessop himself does not go much further than Stoker's notion of a 'certain correspondence'. He refers to a 'structural coupling and co-evolution of changing economies and state forms' and conceives of this occurring in a 'shared inter discursive space' where 'reflexive learning and monitoring of action takes place' (Jessop 1990).

This last and telling comment of Jessop takes us to the place where we might find the intermediary of state and economic relations, the actor involved in constituting these processes and relations The actor, as mediating both 'structure' and 'process', is 'not in a simple sense captive of their role, official purpose or established procedures of the organisation, the participants fill the role with unique content' (Poulantzas 1975). Although there is a discussion here about whether actors are simply bearers of social relations as Poulantzas would suggest, I would not necessarily accept that the two concepts contradict each other. Benson was correct in asserting that

> 'the process of social construction, proceeds through the mediation of interests. Participants' perspectives are affected by the present structure of advantages and disadvantages built into the system' (Benson 1982).

Benson suggests that space for political action and change does exist within the modern state precisely because it is multi-levelled and multi faceted and riven by contradiction. It is this dialectical theory which Benson suggests is the 'explanation of the process involved in the production and reproduction and destruction of particular organisational forms'.

Within this formula the decision rules and the design of policy processes are shaped by the power struggles over the design as well as the content of policymaking processes. If we look at the design of process by drawing on a bottom up methodology or strategic mapping of the implementation field, we can detect the sorts of interests that are influencing decision rules and the dimensions of the struggle over those rules. By further looking at the potential of the institutions and process form for dealing with conflict in the state we can perceive how closure and access routes evolve and by finally dealing with the questions about the meaning of and the strategies which pertain to interorganisation methodologies in the state we can deduce pathways of closure and access.

This suggests that the policy process is not a 'thing' but a series of relations, in fact I specifically argue that process form is both the medium and the outcome of class, ideological and sectional interests. All of these create the nature of process and process form, in itself, acts as an accommodation or constraint on antagonistic relationships. As Healey says, the form of policy process is not

> 'the determined product of dominant power groups but is the outcome of established political and organisational practices' (Healey and Barrett 1990).

Within the context of these debates there is happily a move away from the rational and comprehensive techniques as well as the consensual methods inherent in the incremental or partisan mutual adjustment models. However, as at the outset I made a link between the rise of rational techniques and the period of Fordism and Fordist methods of production and the Keynesian state form, so here I would like to express a link between the decline of these models and the decline in the Keynesian state. In other words, a move to the more radical Marxist and neo liberal models which is echoed in the transition in macro events.

Changing Function and Form of Government

How could this be translated into an examination of the current socio political environment? Fordist politics supported expanding the infrastructure base from which capitalist production could flourish to underpin the growth of mass consumption. A hierarchical structure of state relations descended from nation state level, through subregional or county and local district level. Under the Keynesian Welfare state this structure facilitated redistribution through collective consumption strategies, evening out regional disparities and managing an uneven rate of growth and decline. However although the Keynesian configuration has been termed Fordist, it is to be seen equally as a result of a continuing crisis of over accumulation and an over reliance on accumulation strategies within the secondary and tertiary circuits of capital.

This configuration has altered substantially in the last thirty years. Within this period there have been further shifts towards the centralisation of the state but simultaneously the state form has become more diffuse. Local government, until the mid 1970s, grew in both functional and financial terms but latterly, during the 1980s, it has splintered and the significance of quasi and non elected local agencies, has grown. The vehicles of collective consumption and reproduction are no longer of the same form. Now local governments have a different function.

The economic context for this transformation is the flexibility of multi national capital, the globalisation of production and consumption markets and increased geographical mobility to pursue strategies that will decrease turnover time and limit production costs. Inter-urban competition is a means of fuelling the innovative and accumulating capacity of capital and it is important to local governments which no longer benefit from the redistributive capacity of the nation state to even out the consequences of uneven development. Indeed competition between governments is a key strategy in creating the unevenness which the further accumulation of capital requires. Although the devaluation in the currency of the nation state has perhaps been over-emphasised, it is undoubtedly true that the position of the nation state and local governments within the context of the globalisation of capital movements has altered in the present period.

Yet central government still plays a crucial role in the funding of the capacity of local governments to offer incentives to capital. Until the late seventies quangos tended to be identified as consensual corporatist bodies, involving a limited number of partners. Despite a Conservative commitment to reducing the number of quangos, they have continued to proliferate. The result is that quangos have developed across a number of functional areas. There has been a particular growth in quangos that deal with economic development; in this arena Training and Enterprise Councils, and Urban Development Corporations are pre-eminent examples. The Urban Development Corporation gives central government direct control over the financial resources available for this activity, but simultaneously it has encouraged local fractions of capital and capital generally to participate in the development of the built environment.

So while local governments' redistributive role is declining, indicating the increasing crisis prone nature of western cities and nation states, the wider field of local governance responsibility for the economic development function is expanding. Thus the nation state maintains a decisive hold upon the decision-making procedures of local accumulation strategies but at the same time diminishes the hold of oppositional local governments and its own financial contribution to those strategies. In fact in order to redirect the circulation of capital the protectors of the status quo must be disarmed.

Even though progressive administrations have broken through the strategies of the predominantly Keynesian alliances in the form of new urban left groupings, they have to bend to the coercive laws of competition:

> 'since increasing geographical mobility and rapidly changing technologies have rendered many forms of production highly suspect, so the production of those kinds of services are highly localised, characterised by rapid if not instantaneous turnover time as the most stable basis for entrepreneurial endeavour' (Harvey 1989a).

The quick fixes are emblematic of the UDC experiment and have been crucial in the place marketing campaigns and re-imaging of previously industrial locations. The urban regions have assumed the responsibility for experiments in the built environment of the locality so that local rather than national forces are associated with the entrepreneurial strategies which are to be developed along with the promotion of the role of the local within the international order.

This suggests that a straightforward 'managerial approach' of urban governments popular in the 1960's need to be replaced by a more wide ranging 'entrepreneurial approach' in the 1990's. The fiscal and resulting legitimation crisis of the state (Habermas 1973) has apparently resulted in the inappropriateness of the politics of the Keynesian Welfare state. Though it is yet unclear exactly what new demands are being made on the state form, some semblance of the necessary elements of a functional urban governance already exist.

There is an emphasis on local competition which rejects the redistribution and collective consumption models of the Keynesian state management (Cerny 1989). The development of dual cities within which 'two growth sectors' need to be managed is clearly well advanced – the high technology, information finance and professional sector on one hand, and the low paying subcontract low skill sector on the other (Meyer 1985). These sectors have an interactive and dynamic relationship, but they cannot exist without an intermediary body which will 'rationalise the irrationalities that they exist upon'. How is the growing divergence and necessary polarisation in cities then legitimated within the urban region in the light of dual economies? And if it is not redistribution and collective consumption, then what is the role of the local state? Not only does urban governance have the role of attracting mobile investment into its urban locations as opposed to anybody else's, but it has the additional task of managing 'the dual city' evening out, within its space, the antagonism created by the two sectors.

The prospect of the polarised city is created as a result of the withdrawal of state finance in redistributive functions. It is selective activity on the built environment which will assist the locality in its competitive activities. One way of dealing with this problem is to spread the economic development responsibilities across a number of agencies, across the public and private sectors. Another is to separate economic development from other competing reproduction functions in the local state.

Struggle Over the Change in Function and Form

Harvey portrays the local state almost entirely in spatial terms, in which local relations are articulated politically. This box of coagulative and contradictory forces is able, through 'political processes', to transmit 'demands from urban state levels and back again'. While he acknowledges their complexity, he does not attempt to unravel the processes at work, emphasising instead oscillations between the work place or the point of production and civil society.

The job of the nation state is to 'impose a surface veneer of uniformity' over such subterranean struggle. However, Harvey's account, ascribes flexibility to the 'political space', which houses the urban class alliance. He seems, at first glance, to locate that space within the context of the state, and the state in an increasingly semi or autonomous position in relation to society. I want to suggest that the 'crisis' has created a political space in the locality and within the nation state, where opposing strategies have emerged to deal with the post crisis context for accumulation.

I want to argue that central government has attempted to manage this process by wresting control from local corporate managements and at the same time removing manoeuvurability from local governments. But simply disrupting the existing corporate power structure without putting anything in its place would cause too much disquiet.

Hence the creation of the Development Corporation. It is therefore important to note that although there has been a focus upon the outputs of Corporation activity as an indication of its position vis a vis interest groups and classes, we must analyse the struggle that is taking place on two levels. Indeed the ostensible struggle over benefit in terms of homes built or jobs created can only be seen alongside a more powerful struggle which is taking place over the forms of policy-making.

How does this connect with the concept of structured coherence? If the policy process is the language through which social relations are articulated, and if urban based ruling class alliances are a key element in maintaining structured coherence, then the shaping of the coherence is not just about providing the material conditions for accumulation, nor is it simply about technological and social change. The shaping of the coherence through the various modes of action between the elements is itself an expression of class struggle. Therefore the parameters that many critiques have established for the debate on the UDC, the central local dichotomy and the elected versus non elected state actually articulate class antagonisms. The management of overaccumulation in the capitalist mode of production through the method of 'spatial fix' should in contemporary circumstances, be accompanied by the concept of the 'process fix'. This is the means through which the relations tracing and underpinning production and consumption arrive at a temporary equilibrium, regenerate and reform as transitions in the regime of accumulation are accompanied by changes in the 'mode of regulation'.

Thus we can contextualise the activities of politicians and professionals in the period of 'Fordism' as cohesive, concerned with redistribution strategies and managing interaction between the local and the central. Movement in and out of the coalition seemed relatively easy, though clearly there was some movement between dominant fractions. But there has been a shift in the way in which groups constitute themselves as part of local governance, and the role which they perform within their governance. In examining the struggles which have taken place, first over economic development strategies and then through the processes of economic development and planning in the new quangos such as the Urban Development Corporation, we have a means of observing how the activities of politicians and professionals have changed, how competitive rather than cooperative relations typify local governments, how conflictual and territorial rivalries abound and how closed the decision making process is, how single issue rather than collective issues are dealt with in the locality, indeed rather like the development of an internal market for government funds. This is how I conceptualize the relationship between process form, capital accumulation and the structural edifice of the state. In fact, the struggles which have taken place over decision rules have been significant in shaping what is seen or not seen as legitimate in state activity.

The Job of the Ruling Class Alliance in a Transitional Period: The Development of New Urban Elites

Harvey's model stresses that urban alliances can be both pro and anti growth, and he illustrates this by contrasting the Paris Commune with Mayor Daley of Chicago and the GLC with Baltimore's Mayor Shaefer. This conception of the 'growth coalition' is tied to the global process of capital accumulation, but within the locale it is clearly a vehicle that is shaped by class antagonisms. This view of alliance rationales assists in the examination of the relationship of the strategies of the urban left in British cities as much as it does in looking at the Urban Development Corporation. Both appear to be operating under the shadow of the neo liberal state hegemony. It is also a useful means of interpreting the heuristic

Figure 3.1 Decision-making Framework in the Keynesian and Post Keynesian State

Framework	**Local Govt** (Keynesian)	**UDC** (Post Keynesian)
Desired planning framework	Strategic	Flexible responsive
Desired internal organisation	Large departments within large authorities	Small teams working within or external to the organisation
Desired personnel practice	In-house employment	Use of consultants and agencies
Structure of competition	Internal, inter-departmental	External and inter-organisational
Structure of co-ordination and collaboration	Inter-departmental	Interagency
Desired interagency interface	Overall framework strategic and hierarchical	Fluid ad hoc as and when meetings

contortions of local/central state relations and their interaction with 'new social movements' during the 1980's (Harvey 1985a)

> 'A ruling Coalition in effect speculates on the production of the preconditions for accumulation; it collectivises risks through finance capital and the state'.

The 'growth coalition' or urban alliance therefore exists to speculate, to accumulate, and also to innovate as well as to consolidate. It is driven by a contradiction. But it is there to protect investment and to promote new growth and innovation in order to secure the capital accumulation process. This is a role that is crucial to understanding how alliances act in transitional periods and strategic moments. Harvey suggests that recent trends have suggested fragmentation within the urban region. He sees this as a problem for the urban region and part of the job of the coalition is to provide coherence in the face of multifarious divisions and fragmentation. Although Harvey's account of growth

Figure 3.2 Decision-making Styles in the Keynesian and Post Keynesian State

Style	**Local Govt** (Keynesian)	**UDC** (Post Keynesian)
Dominant value	Public Good	Economic Good
Dominant ethic (Actors)	Professional	Entrepreneurial
Dominant skills	Managerial	Brokerage, facilitation, negotiation
Attitude to competition	Consensus within/ between state institutions and interests seen as dominant	Competition between state institutions and interests seen as dominant
Approach to development	Co-ordinated and planned approach to development	Risk taking speculative approach
Rules of engagement	Participation and action clearly part of democratic process	Rules developed in an *ad hoc* and ill determined way – contingent on power
Legitimation of economic development planning	Legitimation through the state form and democratic mandate	Legitimation through the 3 'Es' and measurable financial criteria
Approach to opposition	Co-option to policy objectives sought – stabilisation and maintenance equilibrium	Delegitimise and exclude dissenting interests from the process
Control of implementation mechanism	Control or delivery of policy	Regulation or delivery of policy
Desired decision making mode	Rational	Incremental and reactive

coalitions' activity is based on class relations of the urban region and providing as it does the foundation for analysing state relations, it raises questions about how we define class in the context of the contemporary state, how precisely we conceptualize state, economy and interstate relationships and how within such a framework we would deal with the micro process of policy relations.

Surprising as it may seem, Harvey established that politics precede economies in the logic of accumulation. Accumulation is the conversion of surplus capital through the combination with the surplus means of production and labour power to create new commodity production. This is seen as speculative activity. But this speculative activity is preceded by another type of speculative activity. Accumulation requires the production of the preconditions of production. Somehow these have to be created, specifically the production of physical and social infrastructure, and there has to be a certain symmetry; prior speculative investment has to match the requirement of future speculative growth. So, in effect, Harvey sees the ruling class alliance as a collective unit ensuring that the risk in the production of such infrastructure is minimised. The tools it uses to do this are finance capital and the state. Like Molotch and Logan (Molotch and Logan 1976), he identifies elements of banking, finance and property capital, construction interests and 'ambitious agents of the state apparatus'. These actors do two things. They obtain profit from the production of the preconditions for speculative activity and they push for or provide the impetus to ensure their future profitability through sustaining future accumulation. Speculation for the preconditions for innovation is heavily dependent on the social and the political makeup of an area because it is heavily bound up with consumption patterns, images and lifestyles.

The struggle to create new coalitions and new elements to the urban class alliance creates meleés and these can in themselves contribute to the 'rational' mix to fit in with the 'winning' combination for a successful urban region. All of this creates a rather dynamic and extremely unstable basis for the alliance. It is trapped into facing in different directions over different issues. It has to be constantly speculative, yet its members will inevitably move towards monopolisation in both economic and productive terms and in the 'modes and politics of government'. It has to discipline its members at the same time as seeking innovation. A way to ensure that a previous 'domain' is broken and produce a melee that is solely based on factions of capital, however, is to 'create' the elements of a new alliance. Therefore, according to Harvey's analysis, there has to be some means of legitimating and implementing this set of relations in order to realise the actual goal of accumulation. There has to be a set of procedures through which this takes place and a form through which the procedures are bound.

Policies for Transition

Managerialism is the organisational strategy that is associated with corporatist social relations and Fordist methods of production. According to many writers there is now a new role and new strategies for local governments to engage in (Jessop 1994).

> 'How the organisation of local growth machines and ruling class alliances that engage in the collective speculation to attract other forms of capitalist development or actively defend a local economy against devaluation, defines what local politics is about' (Harvey 1989).

They suggest that there is something categorically different about the present organisational strategies in governance and it has been called entrepreneurialism. In summary, a definition of an entrepreneurial strategy goes something like this:

> 'It is based on a public private sector partnership, focuses on economic development with a speculative construction of place rather than the amelioration of conditions within a specific territory' (Harvey 1989).

Harvey develops four major strategies within an entrepreneurial paradigm. Competition within the International Division of Labour, involving strengthening the economic base to allow the research and development of new product technology. Competition within the Spatial Division of Consumption, involving upgrading the physical environment, consumer attractions, face-lifts, festivals etc. Acquisition of Key Command and Control Functions. High finance, government or information, cable modern telecommunications facilities, Emphasis on educational and research establishment, good rail road infrastructures, headquarters of banks, government departments etc. Competition over the Redistribution of Surpluses, i.e. military and defence establishments pouring money into big regional offices for sorting out the welfare state.

Jessop uses a different approach to categorise similar developments. As I stated above Jessop has identified three specific types of 'Schumpeterian work-fare state' strategy. The neo liberal strategy includes privatisation, the liberalisation and commercialisation of the residual state sector, deregulation in the private sector and passive state support for market solutions. The Neo Corporatist strategy is interested in providing the impetus to innovation and structural competitiveness, reflecting the heterogeneity of labour markets and ensuring that policy arenas are flexible. Finally Jessop's neo statist solutions include a state guided approach to economic organisation through intervention from outside and above market mechanisms. Here however there would be no consultation with market mechanisms. The state would also set strategic targets for the achievement of flexible accumulation (Jessop 1994). In common with

Harvey he stresses that no one of these strategies is mutually exclusive and that they are used across and between a wide range of operational levels. However the next step in evaluating the usefulness of these categorisation (which are clearly echoed in my own work), is to apply them to particular situations and particular state organisations, which is what the next stage of the book is about.

A Typology of Decision Frameworks and Styles

What I have attempted to do so far is outline the changes in the structuring forces of cities as they relate to our conceptual framework and the UDCs. Now, in turning to the decision frameworks and decision-making styles in the period of transformation, I will attempt to develop a systematic typology of the differences between the ideal operational environments for local governance in the Keynesian and post Keynesian state form and suggest how these might be used to evaluate the different levels and rates of actual change in policy processes across urban regions.

Why is such a typology necessary? There are three reasons for approaching the observation of the UDC in the context of theories of decision-making. Firstly, Harvey only latterly developed the concept of the 'entrepreneurial state' as short hand for a post Keynesian state and while he has concentrated on 'strategies' he has not really looked at policy frameworks or styles (Harvey 1989). With the exception of Painter (Painter 1992) and Stoker (1989b), other writers who have used the concept emphasise entrepreneurial 'strategies' rather than 'post Keynesian' or 'transitional' policy processes. The same problem can be seen in Jessop's neo Schumpeterian workfare state strategies.

This leaves us working on the assumption that strategies that are evocative of 'entrepreneurialism' or the 'Schumpeterian workfare state' are used by organisations that have moved towards a post keynesian state form. Even more misleading is the simple assumption that 'entrepreneurial strategies' in particular are contingent on post fordist practices.Within the political context, entrepreneurialism has been rhetoricised or expressed through institutional collocation. It appears that we are entering something that approximates to a post Keynesian state, taking institutional change, as emblematic of such a form. For example, the fragmentation of agencies, the First Steps approach in the Civil Service, Compulsory Competitive Tendering in local authorities and the internal market in the Health Service.

Interpreting the institutional design of the Thatcher experiments in urban policy of which the Urban Development Corporation is but a flagship, in this way is very attractive. But how accurate is it? How deeply embedded is the trend and what does this language tell us about the processes which the state relies upon in different periods of late capitalism? It is necessary to examine the nature of processual evolution within this new environment in order to develop a complete picture of the changes that are actually emerging and to evaluate fully whether we are seeing a transformation in the social relations which permeate the state, what the meaning of this is and how it is reflected in terms of entrepreneurialism and post Keynesian. We do not as yet possess a typology of policy processes in these organisations, so we cannot hope to have a typology of process form in either a transitional moment within the changing 'mode of regulation' or in a fully fledged post Keynesian state. I attempt here to provide such a typology.

Therefore by measuring the impact that coherence might theoretically have on the shape of processes and institutions, I am developing a framework which begins to look at the rate of transformation in policy styles and frameworks as both the state and the region cope with the impact of major crisis. Secondly I look at the factors which affect the implementation of new characteristics. Finally, it is important to look at the processes of interaction between actors and the decision structure within the UDC, because of gaps which exist in the contemporary studies of Corporation policy. There has been a tendency in analysis of Development Corporations to concentrate on policy content as a means to evaluate the organisation. The interrogation of process can have enormous impact on such questions as why are we achieving certain results in certain areas? What factors are affecting the decision-making process and how does this affect the ability to do the job? Contemporary studies have also been marked by a tendency to assume that the way in which the organisation approaches implementation is simply an echo of its organisational or institutional structure. Or, possibly worse, an instrumentalist approach to assume that the rules, resources and ideas within the UDC reflect those of the central state, as the organisation is simply an extension of that state.

I will look at two sets of factors, the elements of decision making style and the elements of a decision making framework. This methodology benefits from the work of Benson and of Elmore and Hjern and Porter on the value of implementation analysis (Elmore 1985, Hjern and Porter 1981 and Benson 1982). They stress the importance of beginning an analysis of process by focussing on networks of actors, their goals and strategies, in order to assess the impact of local regional and national actors in the implementation of policy programmes.

I will suggest that the decision framework will be used as a means of conceptualizing the influence of structural factors upon the decision making structure and interaction of the UDC. This provides the analytical infrastructure with which I can probe how national legislation meets organisational imperatives and economic demands. The typology of decision making styles which I have developed is the point at which structure meets agency. Decision-making styles are a result of individual actors marrying with the context around them and bringing with them a reflexivity which impacts on the organisational form that develops. The interaction of these sets of factors allows us to explore the antagonisms that are embedded in the regional social and economic infrastructure through the actions of the actors themselves.

Conclusion

I will attempt to operationalise my examination of the case studies on two levels. Firstly I will examine structured coherence within the urban regions according to the dimensions therein; drawing out the technological mix, organisational form and social relations of the urban regions. Having provided the overall context for analysing state interaction with the urban region I will then categorise the regions in which the UDCs have been placed by detailing the 'dimensions of process form' in order to make regional comparisons of the nature of local policy processes. Though these comparisons cannot be said to be scientifically rigorous, it seems to me that in this way we can achieve an adequate approximation of the levels of social and economic relationships within the regional state nexus which will be useful to both the policy analyst and the social scientist.

This second stage in the development of a framework will be important in analysing the agencies of the state. This could then breach the relationship between the macro economic processes of capitalist development and the micro processes of state policy-making. My own emphasis has been to provide such a model within the given parameters of Keynesian and post Keynesian state forms. This then begins to look at the specific detail of the impact of changes in the 'mode of regulation' upon the styles and frameworks of state policy making, thus it will be possible to develop a typology of processes for the post Keynesian state.

There are two notes of caution that should be sounded within such a model, however. The first is that although a regulation approach is a useful theoretical underpinning to a categorisation in the stages of development within the capitalist mode of production, like many others I cannot be categorical in assuming the transformation in regimes of accumulation. It seems to me that when overlaying Harvey's map of structured coherence on the urban region it is unclear if we have undergone a transition to a new regime of accumulation. At first glance it is difficult to estimate whether the inconsistencies which are now being observed within and without nation states are as contingent upon the urban region's structured coherence and regional class alliance as they appear to be, or if they are simply different moments in a transition. However I feel sufficiently confident to say that although it appears to me that we cannot argue conclusively that we are beyond a transitional stage, structured coherence could exert a significant impact on the policy development of an organisation. Therefore my treatment of the regime of accumulation is to some extent incomplete. I have deliberately sidestepped the minutiae of regimes of accumulation, instead concentrating upon the changes in 'modes of regulation'.

I am arguing that there is a change taking place in the regime of accumulation within the global circuit of capital, a successful regime is predicated on a compatible mode of regulation of which the state form is of crucial importance, thus there is a time of experimentation taking place in the search for a suitable mode of regulation. This search in England at least has thrown the nation state into a conflictual encounter with the spatial reach of structured coherence and in effect the regional class alliance (it must be said that experimentation was already taking place at local level anyhow). This conflict has affected the form which the UDCs have developed and the strategies that they have developed, and ultimately their processes of policymaking. At this stage we going to look at the meso level effects before moving on to look at process of policy from which we can deduce, the nature of coherence locally, the nature of the class alliances, the process of struggle which has taken place and the forms of Fordist and post Fordist ways of doing things. These would be used as the framework in which we can observe the operation of structured coherence and urban class alliance in developing the policies and processes of the UDC. This allows us to map out the context of the story. The story of the UDC assumes different time frames not simply as a result of different implementation periods but also because of the different social relations which have had an impact on their conflicts, interactions and negotiations.

4 Bristol Development Corporation

Introduction

Bristol UDC was announced in December 1987. It was one of three 'mini' Corporations, but unlike its counterparts in the North of England, with whom it shared many characteristics, there was a year long delay before it was designated. The delay was caused by Bristol City Council's decision to petition against designation. This, and the parameters of the struggle between central government and the City of Bristol have had a significant impact on the Bristol Development Corporation's ability to implement its objectives and develop local political and administrative effectiveness.

Announcement of Designation

The Secretary of State, Nicholas Ridley, announced the Bristol Urban Development Corporation in December 1987. It was perhaps the most surprising of choices for a UDC as with the exception of neighbouring St Pauls, it exhibited none of the characteristics of inner city decline. The Secretary of State said:

> 'I have chosen these areas because they all have significant amounts of derelict and disused land and vacant buildings' (Bristol City Council 1988).

Earlier in the year David Trippier, Environment Under Secretary, had been more illuminating at a CIPFA conference;

> the new corporations "will create the business-friendly environment and single minded approach but they will be located in areas where existing agencies have not been effective for one reason or another. In some cases there is clearly a hostility to the private sector. It should be possible in most cases to effect regeneration with a relatively modest public sector contribution" (Bristol City Council 1988).

Either the Under Secretary misunderstood the development market or he had an interest in belittling public sector involvement in regeneration. Private sector development has been traditionally reliant on public sector financial injections to prime development and sustain stability in the post war period, and remains so today. As Christopher Thomas, the then Chairman designate of the Bristol Development Corporation, later said:

> 'there was a theory that you could have a cashless UDC, that woûld never have worked').

Corporations could not be established without a massive public sector contribution and by 1987 there was evidence for this in both the adequately leveraged LDDC and in MDC where private leverage was poorly indicated. The impending designation did not surprise Bristol City Council. Like Sheffield and Birmingham City Councils, they had already attempted to prevent a designation. The Chair of Planning felt:

> 'We needed to accept the Lloyds development (on another prime site in Bristol) word got around from the DOE, a message came from the centre saying that we'd better be careful about it (planning permission and planning gain agreements for Lloyds and Cannons Marsh) because Bristol's name was somewhere on a list').

Following the announcement in January 1988, a planning consultancy, ECOTEC was commissioned to carry out the designation study. It became clear in February that there would be opposition to the plans and that the introduction of the UDC would not be as clear-cut a negotiation as elsewhere. This was primarily a result of the activities of a rainbow coalition of urban left Labour party members and professional amenity groups taking advantage of the incoherence of the local political infrastructure.

The first major report on the UDCs, went to the City Council in February 1988. It outlined the proposed UDA but also examined the politics of the UDCs in general. (Bristol City Council 1988). The Council later petitioned against the UDC and began a nine month battle with Westminster, which ended in a House of Lords hearing. It began appropriately enough in the local papers, the Evening Post in early February, in an article entitled 'Battle ground: Who has the right formula to transform Bristol's wasteland?' The protagonists were Nicholas Ridley, Secretary of State, – "the city council has had forty years and hasn't done it" and Andrew May, the Chair of the Council's Planning Committee – "you don't need a government takeover of the area to bring about regeneration" (*Bristol Evening Post* 11/2/88.)

In March 1988 Christopher Thomas, a previous High Sheriff of Avon, and commercial businessman was announced as Chair Designate of the UDC. During the following six months politicians and officials met twice in an attempt to resolve the political stalemate over the establishment of the UDC. David Trippier's first visit was in April 1988. In May 1988 the Council agreed to register its objection to the UDC. In June, David Trippier paid a second visit but could not persuade the council to agree to the designation (*Bristol Evening Post* 29/6/88). It seems that the council officers and members took advice from colleagues in Sheffield City Council on what tactics to employ against central government. Bristol City Council petitioned the House of Lords in June 1988. The Hybrid Instruments Committee conceded that there was a legitimate case to hold an investigation into the designation and established a select committee (Bristol City Council 1988).

The original UDA occupied 420 hectares of central Bristol. It was bounded to the North by the M32 which leads onto the M4; to the West it was enclosed by Temple Way which separates it from the City Centre, and to the South it is flush with the A4 Bath Road. The area stretched from the city centre along the river and the Avon Feeder Canal, up the Avon Valley where it finishes at the St. Annes site in the east (ECOTEC 1988). The central part of the UDA comprised St. Philips Marsh, the central industrial area of the city, housing dirty industry, engineering, chemicals, and distribution uses. The site at the west end of the UDA includes Temple Meads station and goods depot, while, as it moves eastward, there is more open space as well as the derelict area of St. Annes.

The reasons for the Bristol designation were given by the DOE as:

1. fragmented land ownership within the UDA.
2. a problem of accessibility within the proposed UDA, an inadequate road pattern and inappropriate rail and waterway crossings.
3. physical difficulties with the land, including contamination at Crews Hole.
4. one of the poorest environments in Bristol (Parker and Oatley 1989).

But Bristol City Council objected to this reasoning stating that the area designated was already economically successful, that housing and redevelopment were already in progress, that all areas within the UDA had existing plans for development and a good number had planning permission and that there was no serious problem with contamination – only two out of the twelve sites were contaminated, Crews Hole and the former Distillers site (1).

ECOTEC countered, arguing that the majority of the economic inner city initiatives were concentrated in the deprived inner city wards of St. Pauls. Therefore development opportunities were being missed:

> 'Bristol is a highly attractive location for service companies moving from the South East...currently there are no major opportunities available for companies of this kind...the availability of sites is extremely limited'. 'Within the Bristol sub region there is undoubtedly capacity to accommodate a substantial increase in activity. Whilst there is therefore ample supply of high quality peripheral sites, there is a shortage of industrial sites within the city centre' (ECOTEC 1988).

The House of Lords agreed to an amended Urban Development Area on November 10 1988. Lord Dean of Beswick, a Labour Peer on the committee, called for the government to 'extend the hand of reconciliation to City Council by appointing Labour Councillors to the UDC' (*Bristol Evening Post* 20/12/88). To some extent this had already been done by the committee itself, as it had agreed with some of the statements made by the City Council about some of their key sites. The Totterdown site, Old Market and Newfoundland Road and Victoria Street to Temple Meads and Birchwood Lodge to Cannon Farm were removed. This was hailed as a significant victory by some members of the City Council.

The UDC was established in January 1989. It housed a population of 1,000 and employment of approximately 16,000 at designation. The original financial allocation for the UDC was £15 million. However, this has already been exceeded by a further £7 million allocation in 1991-92 and a combined DOE/DT and private sector project of £50 million for the transport flagship, the new Spine Road project. Within the UDA there are Five Flagship projects of which two, Temple Meads and Kingsley Village and the Spine Road, are most significant, have gone to public enquiry and have proved politically contentious. Expenditure on projects other than the flagships is comparatively limited.

Bristol undoubtedly spends the lowest proportion of its allocation on social and community developments. Only marketing has a high profile because of the emphasis which the UDC places on its European strategy. The Development Corporation's involvement in training and employment activities relates only to the business that operates in the UDA. The UDC has established a business club for the businesses that remain in the UDA and has carried out surveys into local employment.

The political culture of the local state in Bristol is historically non interventionist. Instead of a municipal and strategic approach such as we shall see in Newcastle, the local economy has been at the centre of struggles by a succession of local rivalries. Networks of non elected decision-makers have, for the past two hundred years, influenced the strategic decisions of Bristol's local economy and the evidence is, that they still do so (Lovering 1983). Two local authorities were responsible for the UDA, prior to designation – Bristol City, for Planning and Housing and Avon County Council in charge of highways and strategic development. Bristol City Council was Labour controlled at the time and Avon was a 'hung' Council. Both had experienced instability in political control throughout the 1970s and 1980s.

As I suggested earlier, the local authority departments which are most likely to be affected by the remit of the UDC are the Planning, Economic Development and Estates departments. Within the City Council, Planning was the most significant department. Its ethos in the post war era was pro-growth, but also what is anecdotally talked about in Bristol as a 'Merchant Venturers tradition'(*Venue* June 1990). By this it was meant that the department adopted a hands-off approach to development, especially commercial development, leaving commercial capital as the de facto planning authority for the city.

In 1975 a new Director of Planning was appointed and developed a highly interventionist planning process. In his opinion, the city centre had been ruinously scarred by the pro-growth office development of the 1950s and 60s. His interests in design coalesced with those of the strongly articulate design and consumption amenity groups who were viewed as anti growth coalitions, presaging the subversive green movements of the 1980s. These groups comprised predominantly public sector professionals and with their backing the Department became more consciously regulatory and more insistent on its strategic planning role.

An early internal coup against the changing department failed and the poor relationship between Planning and the rest of the authority continued after the appointment of yet another new director, Ian Patterson in 1977. This was, in part, due to Patterson's political demeanour, and his inability to tow the Chief Officer's line, but it was substantially influenced by the changing political sensitivities within the authority and its movement to the left (*Interview* no. 4). In addition to the internal wrangling within the authority, the loss of transport and engineering powers to the County after reorganisation in 1974 made infrastructure planning more difficult. The divide between the two authorities manifested itself in the championing of separate causes – the County promoted roads and the City promoted conservation and rehabilitation of buildings.

Meanwhile the Planning department faced intense criticism for its anti development bias. In 1985 local property interests and the Chamber of Commerce issued a report 'Planning Difficulties' attacking the Department for its anti growth stance (*Interview* no. 5).

As I have said Bristol functions at a political level through a range of informal networks founded on historical associations. The major contemporary political challenge to this model of municipal operation has come in the form of a new urban left, which occupied local public office in the late seventies and the early eighties, and which was determined to challenge the commercial hegemony and paternalistic decision-making style of the established networks. Outside the institutional environment, a multivarious selection of environmentalist, conservationist and social 'urban movements' echoed the themes that were explored inside the Town Hall.

At the economic level, the increasing dissociation between headquarters of industrial and commercial Capital and the locality similarly undermined the status quo. In political terms, by 1979, a transformation had taken place in the City Council. There was an influx of new young urban left councillors, often professional, many of whom had an background in community activism. Within the Planning department this created a more politically conscious and territorially expansionist department causing some friction within the authority as a whole. By 1984 the left had consolidated an uneasy but significant hold within the group and took over key chairmanships. They became imbued with the new leftist strategies that were infiltrating the GLC, Sheffield and Manchester, and a number of other Councils (Wainwright 1989).

Urban Policy in Bristol

Although there has been a gradual transition in the Bristol labour market during the last thirty years, the diversification of the local economy, especially in the post war era, and the spatial differentiation therein, meant that the economic crisis of 1973 did not have the same effect in Bristol as in other major UK cities, though the collapse in the property market at that time was a significant and traumatic event. Similarly the state fiscal crisis did not impact to the same extent as other local authorities in Bristol, because of its traditional non municipalist emphasis, and any effects that were felt, came slightly later than elsewhere in the UK.

The aerospace and defence industries did not experience the contraction of many other manufacturing industries until the late 1980s. This happened at the same time as deregulation of the London finance market and 'black Wednesday' and conspired to undermine the local property and financial services sector. Consequently the urban policy agenda in Bristol did not assume the cohesiveness of other urban centres and its internal organisation remained poorly equipped to garner urban policy funding from central government. Only latterly did Bristol's statutory and voluntary sector really begin to come to terms with the impact that a major economic restructuring would have on the locality; and because of the nature of local industry and the internal socio-economic divisions within the City, its response to these circumstances was diffuse and contradictory.

Bristol was unable for some considerable time and only as a result of special circumstances able to attract urban aid. However this is not to say there was no urban policy in the city, an Economic Development Board was established as early as 1976. This was an all council group and included employers, trades unions and educational establishments in the city. In the same year the City Council established an Economic Development Subcommittee. The policies of the committee were to protect existing jobs, maintain an adequate supply of land, attract inward investment and support for small firms, thus the strategies were growth orientated. They developed a three pronged approach establishing the new harbour, attracting primarily North American inward investment and building enterprise workshop units. By the mid eighties in line with the political changes which were taking place the Council moved towards social equity strategies. The St Pauls riots of 1981, highlighted the discrepancies between the black and white communities and the discrepancies in wealth which existed in the city. The Council started talking about guiding inward investment into targeted areas of high unemployment and towards vulnerable groups and a series of local economic initiatives was established.

The local authority was influenced by three factors in its economic policy from 1984 onwards:

a) desire to expand economic development work beyond promotion and marketing;
b) increasing restrictions on local authority finance;
c) the targeting of specific grants from central government and the EC for employment and economic purposes.

This approach was only really consolidated in 1986 when 'A socialist strategy' was adopted by the District Labour Party. This laid down policy guidance for the ruling Labour group. It included:

> proposed planning agreements with the private sector to regulate conditions of work, including contract compliance policies ...
>
> expansion of municipal enterprise and the diversification of employment strategies into co-operative and socially owned enterprises (Bristol District Labour Party 1986).

The changes in the outlook of the Councillors and many of their support networks inside the Labour party and beyond in the community movements of the area, penetrated the Planning Department of the City Council, more deeply than the

Economic Development Department. EDU's officers remained firmly committed to pro growth strategies and non interventionism, feeling that the 'local' in local economic development marginalised their impact upon the city's development as a whole. Later, while the Planning Department, assumed an offensive stance towards the Development Corporation, the Economic Development Department was more hesitant, fearful of getting involved in 'sensitive politics' or indeed of displaying opposition to the Planning department's, or their own, Chairperson's views (*Interview* no. 6).

Similarly the County Council did not garner the urban left impetus in the same way as the city had done. The County Council had separate responsibilities for the area as a Highway authority and had strategic economic development and structure plan powers. It had co-operated in the Economic Development Board of the late 1970s and had established an Employment and Economic Development in May 1982 (*Interview* no. 7). But its stance on the UDC prior to designation was uncommitted.

The Politics of Urban Regeneration and the UDC

The Chair of the Planning Committee, who was also the leading 'left winger' in the Council, waged a serious fight over the definition of the boundary of the UDC. The sites of Old Market, Newfoundland Road (which had been extended after the initial announcement) and Victoria Street to Temple Meads area, were of particular concern because of the Council's plans for those sites (*Evening Post* 31/3/88). In May 1988 he made it clear to the press that the inclusion of the site was one the reasons for the objections raised by the City Council and the following month he raised it with David Trippier, who was on a visit to Bristol to negotiate with the local authority. The Minister was forced to acknowledge the case that the council was making for the site:

> 'I have given the City Council assurance about the future of the sites including Bond Street' (*Evening Post* 4/6/88).

As Andrew May was later to confirm:

> 'I am quite sure that the redrawing of the boundaries was quite critical because it removed certain key sites which were in local authority ownership and I think that the initial strategy was based on the expropriation of those sites, they'd just take them, give current use value to the council and the development gain that they would make would have been enormous. The removal of Bond Street, sites in Victoria Street and Totterdown, meant that the sums changed drastically, one set of consultants was sacked and replaced by another. The whole strategic approach foundered'.

The resolution of this issue resulted in a split in the local district Labour party and within the ruling Labour group. The District Secretary and the former MP for Bristol North West, Ron Thomas, stated in the Evening Post he understood why Andrew May and Dawn Primarolo (MP for South Bristol which included Totterdown)' felt they had gained something' but that while they had saved their 'jewels in the crown, we are left with the crown of thorns in Bristol East' (*Bristol Evening Post* 18/8/88).

This disagreement serves to emphasise the fractionalised nature of Bristol Labour politics. The divisions within the Party and the Council, which stemmed not only from different political positions, but also different occupational backgrounds and interests, made it truly remarkable that there was any consistency in the Councils position at all in relation to the UDC.

In July 1988, the same month that Bristol presented its petition to the House of Lords, further negotiations were taking place away from press and party scrutiny. The City Council's Planning Officer, together with the Developer, Mr Peter Weaver, met with the DOE to attempt to retrieve the Victoria Street site which the City Council had been directed earlier to refuse planning permission for a 140 bedroom hotel, 40,000 square feet of offices and 170 car parking spaces. The DOE at that stage intended to object to the planning permission which the City Council had given for this site. The Director of Planning alleged that the DOE had stated that the Victoria Street site 'would be the jewel in the crown of the UDC'. Effectively the DOE was suggesting that the City Council stand back in order to allow the UDC to develop the project. In fact the DOE gave way on this ruling some time after the meeting and the city was allowed to continue with the Development (*Interview* 4).

One reason why land value was such an emotive issue for the actors and the agencies involved in the fight against designation, was the ostentatious speculation over St. Annes Board Mills site, especially by the developer and property speculator, Mr. Peter De Savery. Mr. De Savery had bought the St. Annes Board Mills site at the eastern end of the UDA, in April 1988. After remaining derelict for seven years due to the closure of the Wills Tobacco factory, the site had changed hands three times in nine months. Hanson PLC had sold it to property developers, Mountleigh, who sold it on to De Savery. He sold it four months later to the Fairbriar Group for a profit of £5 million. (*Evening Post* 18/6/88).

The speculation received a lot of press coverage, because of the implications of rising land values within the city and also because of the widespread publicity about the property deals made infamous in the London Dockland site. A left wing Labour councillor, Ann Thomas, called for the Council to CPO the site to protect it from the speculators (*Evening Post* 18/6/88). The De Savery case was not alone in provoking alarm about colossal exchange values. Another of the

Bristol City Councillors was concerned that a local sports ground would be: 'stolen from local people from private developers and speculators...there is a clear choice, public amenity or private profit' (*Evening Post* 30/1/88). The councillor clearly saw himself defending public assets.

The fears of the council leader, Graham Roberston, that areas zoned for industrial and employment uses would be redesignated, cannot have been eased by the words of the Secretary of State, Nicholas Ridley, who stated when he was talking about new uses: 'Once you realise the potential of this area of land it will be able to become much more valuable and make money'. (30/1/88 *Evening Post*) One of the County Councillors representing a ward covered by the UDC said: 'Its main purpose is to make a quick profit for developers – it's all based on unlocking land to hand over to the private sector. For example Mountleigh Properties have already made £6 million out of the Board Mills site' (*Marks; SAUS 1988*). It is important to note that councillors also had different views according to the positions that they occupied within the Council.

Preceding Strategic Framework

There are four planning documents which cover the area, the Avon County Structure Plan, published in 1983, the Bristol Development Plan Review, the East Bristol Draft Local Plan and the City Centre Local Plan. The Structure and the local plan are most pertinent to the UDC and later proved significant in the Public Enquiries over key sites in the UDA. However, it is not just the content of the Plans which are important in the context of the UDC/local authority relationships, but their style of implementation. The City Centre Local Plan involved extensive consultation and public participation in line with the ethos of the Planning Department and the community orientation of its politicians. Although the City Council attempted to involve the UDC in development of the framework, applying it to the relevant sections of the UDA, the UDC was uncooperative, suggesting alternative plans for the designated area, differing from the Council especially on the requirements for office space and density. This in part reflects the different approach to development on the part of the UDC. Council and UDC frameworks for the Avon valley area, and St. Annes proved less controversial and were reasonably similar, though there has been conflict over the residential plans and the Council has objected to the lack of provision for social housing within the UDC framework.

The Urban Development Corporation

Establishing the Board and Management Structure

The Urban Development Corporation's board confirms Lovering's thesis, about the dominance of historical networks operating on contemporary public policy initiatives in Bristol. Although five of the current members of the board are involved in the public sector (including the University and the previous Somerset Health Authority), public sector influence does not appear to be as important in Bristol as elsewhere in England. Lovering hints at the secrecy inherent in the ruling oligarchy of Bristol and suggests this has been accompanied by an anti public sector stance. Again this is borne out in the backgrounds of the personnel involved in the UDC and in their approach to policy implementation.

Lovering was referring in part to the role of the Merchant Venturers Society in Bristol policy domains. The Society of Merchant Venturers was established in 1552 to operate a closed shop in the Port of Bristol, giving their members sole rights to handle the foreign trade coming into the Port. Bristol was for many years in the eighteenth century the second port next to London in England. The society was involved with the triangular trade in slaves, sugar and cotton in the West Indies. As the Docks went into decline in the nineteenth century the Society started to invest in property development. Reluctantly the society admitted entrepreneurs to its ranks, as the industrial revolution spawned a new ruling class in Bristol. The Commercial and Professional classes were important in maintaining the position of the Society in Bristol. The Society still occupies a leading position in the major decision-making structures in Bristol.

This is in part due to the tremendous legacy of influence imparted by their widespread investment in property and in part due to the leading positions that present day Venturers occupy in business and commerce within the City. Venturers comprise 50% of the membership of the Board of the Urban Development Corporation. Indeed one of the property assets which belongs to the Venturers is a large portion of the Bond Street site which was the subject of so much contention in the Development Corporation battle. Christopher Thomas became Chair of the Development Corporation in March 1988. His social and political network was thought to cement in the position of the UDC in the locality, as he stated in an interview the following year:

> 'The family firm which had been going for 200 years (The Sheldon Bush Patent Shot Company) was sold in June 1987 so this job came along at just the right moment'.

Thomas is a member of that 'Bristol magic circle of influential old money, old family network' and is a Venturer. He shares this connection with four other members of the Board and membership with two. The membership gives him direct access to two cabinet ministers, (William Waldgrave and Tom King), as well as an ex Prime Minister (Margaret Thatcher, is the only woman Merchant Venturer) (*Venue* 22/6/88), as well as societal connections to Sir John Wills of Imperial Tobacco, Bristol Water Company, Bristol and West Building Society and the Bristol United Press. Though as Mr Thomas, a former Lord Sheriff of Avon, has directorships in the Bristol Water Company and Bristol and West Building Society, his influence is widespread. He sits on the University Council, together with a number of other Merchant Venturers, and two of his UDC board members (Stella Clark and Philip Gregory). The most recent addition to the Board includes the wife of the late Lord Chancellor and Merchant Venturer, Lord Merrison. Christopher Thomas was the first Chairman of the Somerset Health Authority after reorganisation in 1974. He is also a local director of Hill Samuel, the regional board of the Regional and General and the Western and Provident Society.

His background confirms the societal bias of the Society and its emphasis on the commercial ethos of its members' activities. When he was asked what the most important quality that the Corporation brings to the City, he answered 'it's our commercial approach' (*Interview* 3).

'To become a Venturer a man needs to be prominent in commercial terms in the city' (*Venue* 22/6/88). The Society has extensive links to the Conservative party, through direct membership – William Waldgrave, or indirectly Charles Clarke, (Solicitors Osborne Clarke) who is husband to Stella Clarke, sister of Tom King. Charles Clarke has other connections to Christopher Thomas, as a member of the previous Health Authority. It is also involved in charitable work, especially in education. Although the Venturers of the 19th century opposed the establishment of the University and later the Polytechnic (now the University of the West of England), they now have extensive involvement in both. Now the University has an extensive involvement in the UDC as Christopher Thomas, Stella Clarke and Philip Gregory sit on the University Council and Lady Merrison's husband was the previous Vice Chancellor. Andrew Reid, who in 1992 was made Vice Chairman of the Development Corporation, is a Merchant Venturer and is a former Chairman of Imperial Tobacco, thus shares connections to Western United Press alongside Christopher Thomas. Philip Gregory is regional Secretary of the Trades Union Congress, but shares membership of four other organisations in the area with other UDC board members, the Greater Bristol Trust, on which Stella Clarke sits, amidst half a dozen Merchant Venturers, The Bristol Initiative. on which Christopher Thomas sits, the University Council (above) and until recently the regional health authority, where again there was a cross over with Thomas and Charles Clarke. Timothy Thomas is a Chartered accountant who is a partner in Price Waterhouse in the City as well as a Merchant Venturer, though he does not seem to have any other connections. In 1989 there were three councillors on the Corporation, one Labour and two Conservative members, one of whom was also a chartered accountant. In 1992, the Vice Chairman and ex Lord Mayor of Bristol, Jack Fisk, was replaced by Graham Robertson, the Leader of Bristol City Council, who was, in fact, a full time official of the tobacco workers union before retirement. Other than the public sector representatives, there were two other members of the board who appeared to have no cross linkages in 1989 – Rodger Smedly, an engineer, and David York, a surveyor. Roger Smedly was replaced in 1992.

Like the boards of other Development Corporations, the UDC appears to solidify preexisting networks of influence. But while in many other areas the local authorities and the public sector have been a significant influence, in Bristol this influence resides in more opaque networks, networks whose rules of operation are correspondingly difficult to reveal. One other interesting factor in the Bristol Board however is its lack of allegiance to local manufacturing and industrial interests; it is dominated by commercial and property interests and this is echoed in its policy orientation.

Strategic Framework

The delay in establishing the Bristol Development Corporation, combined latterly with its almost total emphasis on property and infrastructure provision, to the exclusion of labour market, environmental and community initiatives, left the Corporation with only one major realistic regeneration major project capable of realisation. This was the Spine Road. This project attracted minimum leverage and a maximum infusion of capital funds from central government departments. In addition to the limitations placed on its development activity by external economic and political factors, the UDC found itself competing with neighbouring North Avon in attempting to attract prime site development in a dead market, while the local authority found itself in the unusual position of having developed most of its available land as a result of market trends.

The commercial ethos that is represented in the board – there are no representatives of industrial capital – is more or less carried through to the senior staff, and without the public sector balance that occurs in the other UDCs, only one of the executive directors, who has since left the Corporation, has a public sector background, (though the Chief Executive began his accounting career with a local authority before moving to the private sector and then the British Tourist Board) (*Western Daily Press* 2/2/88). The Corporation maintains a small staff, 25 in total, drawing on extensive consultancies. They adopted a comprehensive planning approach on their inception, publishing their draft development strategy three months after coming into existence, 'A Vision for Bristol'. The Corporation originally asked for responses to their 'Vision' document in the middle the local authority recess, highlighting their impatience with traditional practices. They received responses to the circulated document a month later on the 20th and 21st of September from the City and the County Council respectively. Because of the nature of the responses and subsequent reworking of the strategy, the UDC published its definitive development statement in March 1990.

The debate over the strategy cannot really be separated from the coexistent discussions that were taking place about the Code of Practice but additionally, difficulties about the strategic approach of the Corporation, which conflicted with the local structure plans, were also emerging. The Development Corporation claimed later that the difficulties the councils were experiencing were based on 'misunderstandings' of their document and 'mistakes in the strategy diagrams' (Bristol City Council April 1991). In fact the Development Corporation withdrew some of the strategies, especially in relation to the Temple Meads and Kingsley Village and the Central Development Area, including St. Philips Marsh, before circulating the revised strategy document in the summer of 1990.

The revised Strategy was distributed for public consultation, comments were invited from statutory agencies and through a public exhibition of the Corporations plans, from the general public (Bristol City Council 1991). Some of their initial difficulties over their strategy and their relationships, were related to their aggressive approach, to the local private and the public sectors. They were particularly interested in pushing up land values and obtaining planning gain:

> 'it is perfectly legitimate for the landowner to want to make a profit..but he shouldn't have it all, we want to capture...a fair proportion of it to plough back into improving the infrastructure' (*Interview* 3).

The Corporation fought shy of controlling development in theory (though the changing fortunes of the property market made them far more reactive than proactive).

Inter-agency Relationships

The tenor of Bristol Development Corporation's relationships with other agencies or groups in the area was established in the fight between the council and central government. None of its relationships with other agencies operating in the area, could be described as good. Firstly within the local 'community' BDC have been criticised by many for their lack of community development and consultation frameworks (*Venue* 1990). Their lack of commitment to community activities is borne out by the small level of financial assistance that is recorded in their annual reports (Bristol Development Corporation Annual Report 1990). They counter this claim by suggesting that they have carried out extensive surveys on employment and on the needs of businesses, within the UDA (BDC Employment Survey 1990). However, most of this information relates to the organisational requirements of the Corporation in relocating businesses, especially in St. Philips Marsh and around the Spine Road. Two of their attempts to host community activity, one using an ex Sheffield Councillor as a consultant in community development work and another using a London consultancy, have failed (*Interview* 6).

The consultation processes of the UDC have involved very little direct contact with the public. Consultation over their plans has been done via exhibitions and huge mailings of their documents to the relevant public. This type of consultation, together with the limited meetings with the local authority (every six months), has effectively insulated the Corporation from the locality. The local community has responded to the UDC in a fragmented way. A number of residential, amenity and party political groups were initially contacted through a series of meetings that had been organised by the City Council to consult on the UDCs plans. Those who lived in the UDA seemed to have parochial concerns; they were worried about likely occupants of the newly developed area. On the whole 'community' activity of a more generalised political nature was filtered through a series of pressure groups. These ranged from environmental groups like Friends of the Earth to local amenity groups which were concerned with urban design impacts of the UDCs flagship projects. It was these bodies rather than the residential groups which seemed to be more vociferous at the subsequent public enquiries over developments such as the Spine Road.

National economic development networks which had been established in local authorities worked with local constituencies in Bristol to develop a response to government plans for the area. Sheffield City Council which had been designated at the same time as Bristol, and faced the same decision about petitioning, became heavily involved in Bristol's case. Sheffield's Chair of Planning, David Skinner, and the then Chair of Employment, Helen Jackson, visited the City in September of 1989. Both underlined their support for Bristol, but when asked for the reason why they, as a radical left authority, had chosen eventually to accept the UDC and negotiate specific terms with them, they cited their links with the private sector. As David Skinner stated:

> 'In Sheffield we have been able to work with the private sector. Although we have political differences we are able to put them to one side. That does not seem to have happened in Bristol' (*Bristol Evening Post* 21/9/88).

The Evening Post later interviewed the new Chief Executive of the Development Corporation in Sheffield, who echoed David Skinner's argument:

> 'I would not have liked to have come to a post like this knowing that the

> City Council (Sheffield) was totally and utterly opposed to it...Much of the partnership was already beginning before the UDC was proposed...Sheffield's situation was much the same as Bristol's but the partnership between the local council and business is a crucial difference' (*Evening Post* 30/9/88).

But it was not only the public sector who were initially suspicious and latterly cynical about the UDC. Not everybody in the private sector enjoyed a good relationship with the UDC. The property industry and 'old Bristol' were on the whole sanguine, certainly not opposed to the UDC. There were elements of the private sector which expressed doubts; the new private sector partnership, represented by Denis Burn of the Phoenix Initiative stated:

> 'We wrote to the Government saying please recognise we have a partnership here in Bristol... It has been my fear that the arrival of a UDC may cause considerable friction in the city and so will set partnership back a long way' (*Bristol Evening Post* 16/12/88).

The letter from the Chairman of the Phoenix to the Prime Minister objecting to the UDC was published by the local listings magazine and stated that:

> '...it will be a great pity and in my view ill time if Nicholas Ridley declares there is to be an Urban Development Corporation in Bristol. Sir Robert Wall (the leader of the Tories on the City Council) has privately expressed the hope that he may win control of the City Council in May and would wish to run the City along traditional lines. Could you persuade Nicholas Ridley to give Phoenix, the non political private enterprise, a chance to succeed?' (Correspondence between Phoenix and Downing Street).

The private sector did not want central government interfering as long as they could revert to the position where the city ran the council and the private sector ran the town. Andrew May was later to say,

> 'There isn't really a private sector culture in Bristol engaging with the local authority. For whatever reason it goes back a long way, way before I was involved. It doesn't have that kind of corporate edge that these issues have in Newcastle, for example. Bristol comes from another tradition'.

The relative indifference of the private sector carried on after the UDC was established. The CBI and the Phoenix took a long time to be integrated into the structure (*Interview* 4) and it was also to cause problems with the embryonic relationship between the City Council and the Phoenix:

> 'we had some discussion with the Phoenix where they were very positive

with the LA. We thought that might be worth exploring but the inception of the UDC wiped that out' (*Interview* 4).

The Chairman of the Development Corporation, offers perhaps the most simple explanation for the Council's poor relationship with the UDC:

'The City Council understandably didn't like us because we'd been set up to take away their function' (*Interview* 3).

Oatley suggests that the stance taken by the Council generated a lot more division than was publicly visible, particularly in the planning department. There was a lobby to actually 'get our hands dirty, get in there and even put people on the board'. But there were also differences in the Labour group between Andrew May and other members of the traditional leadership in the council and backbenchers who were not on the left of the political spectrum.

The local members of Parliament opposing views apparently corresponded with the dominant interests in their own political party at the time. Dawn Primarolo, Labour MP for Bristol South, saw the issue in terms of accountability and democracy, and attack on the local control apparently marshalled by the Council. Jonathon Sayeed, as a new Conservative member in a recently Labour held seat wanted a regeneration project on his 'patch'. He was later quoted as saying 'I got the UDC, I was responsible, I took photographs of the area and talked to the minister'. At the pre enquiry stage, the hung County Council did not engage in debate to the same extent as the city. Their position noticeably sharpened as they entered negotiations over the Code of Practice and Consultation and when issues of employment were to occur in the labour market structure in St. Philips Marsh as the development of the Spine Road progressed (Interview).

The Code of Practice

The discussion between the UDC and the City Council over the Code of Practice was significant in the UDCs first year of activity. Initially, the Code split the ruling Labour group of the City Council. Discussions over the Code of Practice were taking place as early as January 1989 in the Planning Committee. The committee was using Sheffield City Council's agreement with Sheffield UDC as the foundation for their own Code of Practice (Bristol City Council Minutes 1989). The Code was agreed by all parties of the council but some Labour members were worried that the council might be 'bootlicking the usurpers' (*Evening Post* 19/1/89). The City Council were also disregarding the District Party boycott on dealings with the UDC.

Within the City Council, UDC business is delegated to the Planning

Committee. The leader of the Council and other senior councillors at that point thought that there should be no contact with the UDC, while Andrew May who was Chair of the Planning Committee disagreed, insisting that they should negotiate concessions from the UDC (*Evening Post* 16/1/89). In fact the Council tried to negotiate over the issue of democracy as Sheffield had done, stating:

> 'the example of Sheffield City Council serves to illustrate what can be achieved to democratise what many see as a potentially undemocratic body' (Bristol City Council 1991).

Bristol City Council made a series of demands, some of which were acceded to by the Corporation. Many remained outstanding (2).

In addition to strategic principles, the practical mechanisms for the operation of the relationship between the two organisations was also the subject of negotiation. There are two levels to the policy relationship, strategic, and day to day issues about planning applications. For strategic issues there is a 28 day response time for the City to respond to applications from the BDC. For planning applications the City's Planning and Traffic Committee shadows all the planning applications that are put before the Bristol Development Corporation.

On the 24th of November 1989, the City Council received the Corporation's final draft on the Code of Consultation with outstanding conditions unmet. Between this time and the Council's acceptance of the Code in February 1990 a number of meetings between the city council and the UDC took place. The minutes report forthright views being expressed by both parties. The Council continued to press for open meetings, consultation with the community, and access to the Corporate Plan. While the Corporation stated that they were seeing voluntary and local groups independently, including BCVS and the BCRE, that they had not formally received a copy of Bristol Local Plan so therefore it was difficult to achieve correspondence between there own planning strategy and that of the council. Finally they objected to comments that were made in City Council committees about the Corporation (3).

The discussion over the code of conduct impinged on other issues of policy between the UDC and the City Council:

> 'for a long time the discussion over 'procedure for governance' was a stumbling block to any discussion about strategy' (*Interview* 7).

During the negotiations very few further amendments to the original document on consultation were accepted. In fact, some of the original sections were deleted, such as the monthly note of board meetings. The Council however accepted the terms of the consultation agreement (Minutes Bristol City Council Traffic Committee February 1990). Although the County Council adopted a much lower profile in the initial objections to the UDC, they took a somewhat different

Figure 4.1 Regional Context of the Bristol Development Corporation

Figure 4.2 Bristol Development Corporation UDA

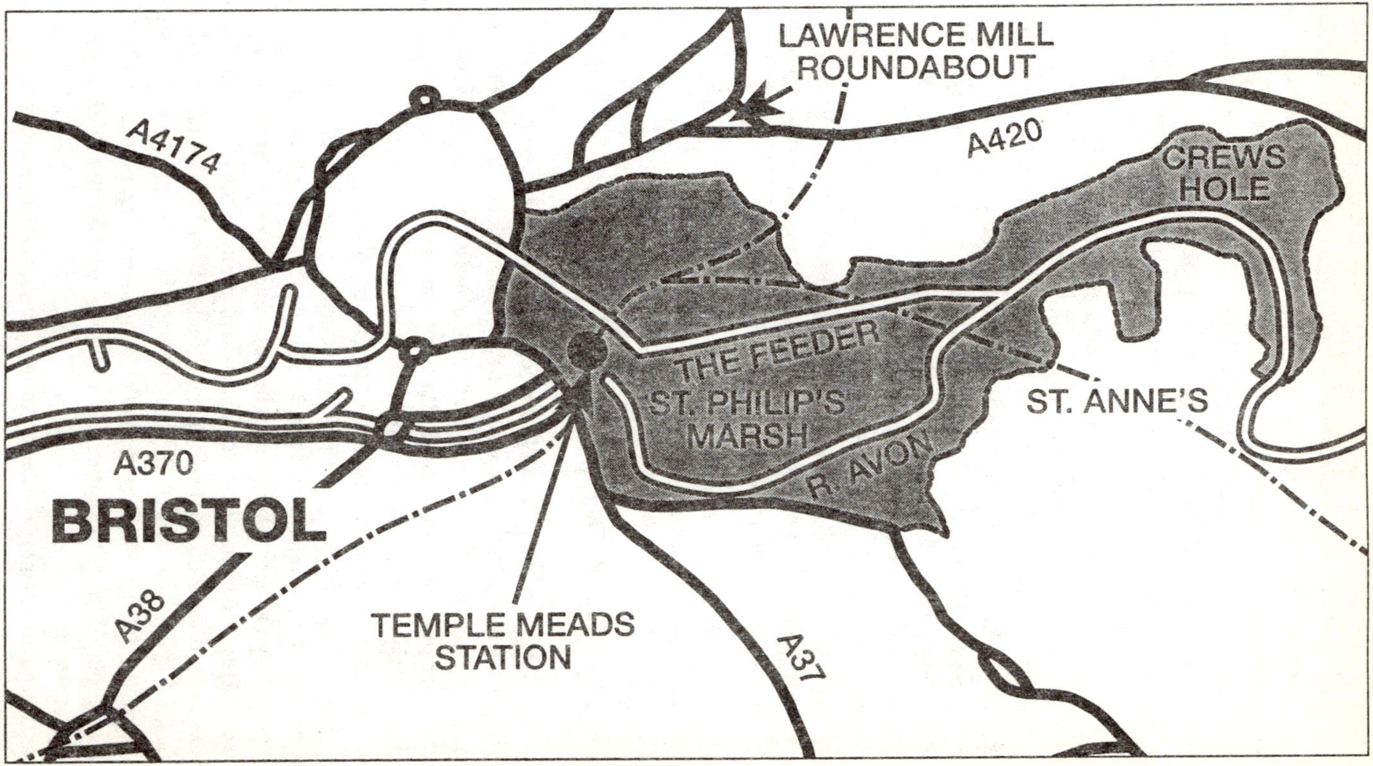

stance in determining a Code of Practice with the UDC. They established a UDC committee and negotiated with the Corporation. The County identified eight principles that they wanted to be put within the Code of Consultation, all of which concerned open and democratic decision-making. The UDC submitted a draft code to ACC in November 1989. The County Council threw it out. A revised Code was submitted to the County Council in January 1990. Although it was still thought inadequate, it was adopted by the UDC without further comment (Bristol City Council 1991).

The UDCs dealings with the regional Department of the Environment shares similar characteristics to their dealings with other agencies in the public sector. Though the UDCs frustration with the regional DOE is the one characteristic that is very similar in the two case study areas. The Corporation perceives both the regional DOE and central government meetings as interference in their work. The Corporation's level of financial delegation is only £30,000 and there is tremendous resentment within the Corporation that they have to report to the DOE, let alone that they have to secure permission to allow projects to go forward. Given this it is perhaps somewhat less surprising that there is some hesitancy in their commitment to work with the local authority. This allows the Chairman to state with ease:

> 'We are undemocratic, we are not accountable to the local people except through the Secretary of State'.

The UDA: Sites and Flagships

The industrial history of the UDA began in the seventeenth century and was significant well into the latter half of the nineteenth century. St Philips remains a centre of small and dirty industry, with housing infrastructure for local workers. Post war development in St. Philips is marked by the displacement of high density housing and a marked increase in speculative development, particularly small industrial units during the last twenty or so years.

Since 1980, 460,000 square feet of industrial units (Punter 1990), 30% of the total industrial development completed in the City, have been constructed in St. Philips. By October 1983 three schemes were under construction, all of which were developed on public sector land with planning permission totals for 700,000 square feet (463,000 on the former St. Annes Board Mills and 155,000 of high technology units) (Bristol City Council 1988). St Philips contains about five hundred small and medium industrial companies. They provide 7,000 jobs. The remainder of the jobs in the area amounted to a further 8,000 at the time of designation.

This employment is concentrated in a number of larger companies, including

British Oxygen, Strachan and Henshaw and Dolan Corrugated Containers, as well as a number of public sector companies, British Rail, the GPO and SWEB. Since 1981, 1,500 manufacturing jobs have been lost in the area. Before designation there were 150 acres of unused land in the UDA. Factory closures, rail land becoming redundant, sites being cleared for commercial development and sites being cleared pending highway works contributed to the rising derelict land bank (ECOTEC 1988).

The Avon Valley

The Avon Valley is at the east of the Development Corporation's area and comprises 130 acres of primarily rural land. It included the previously controversial 70 acre St. Annes Board Mills site, which is prime development area (see above). Although outline planning permission was granted fairly early on granted for a mixed use scheme that will include 650 homes, contrary to the original specification, there are no social units within the area, 100,000 square feet of new business, industrial and commercial space. Another housing area of 250 units was originally planned at the previously contaminated site at Crews Hole across the River from the Board Mills site. The remaining area provides open recreation space and leisure facilities. There has been little controversy about the uses on this site as they correspond to structure plan requirements, even if they do not provide the social housing which the City had advocated for the site. There has been little progress on site and less than a quarter of the houses proposed for the area have been built. The original developers for the area pulled out of the negotiations as the property market slumped.

The Avon Weir

The construction of a Weir in Bathhurst Basin received Parliamentary approval in 1992. The objective of the Weir is to improve the river corridor so that it will provide complementary environments for commercial and tourist development around the river. Although there were some objections to the Weir by the City Council, the passage of the Weir Bill through Parliament has been relatively uncontroversial, though time consuming.

The Central Development Area and the Spine Road

The Central Development area comprises 500 acres of primarily industrial land in the heart of Bristol and at the centre of the UDA. The introduction of the Spine Road, North South through the Central Development Area from Lawrence Hill through to the A4 Bath Road, linking the CDA to the M32, M4 and M5 to the

North of the City and the M4 and A5 to the South, will create (together with the new barrage, the Avon Weir) a new area called St. Philips Island (BDC Annual Report 1991-2). The construction involved the division of the CDA into three distinct development districts. This leaves St. Philips Marsh mostly devoted to its current uses.

The three planning districts are: the South which contains plans for corporate headquarters and commercial campus. The East, the 23 acre Marsh Junction site, now a Corporation joint venture with Hanson PLC, to whom the site previously belonged and with whom the Corporation quickly concluded a negotiation to stave off his objection to the £55 million spine road project (*Estate Times* 14/6/1991). This area will contain mixed development, retail and commercial uses. In the western tip of the site there will be a mix of commercial uses and industry.

Both the County and the City Council have officially objected to the plans for the Spine Road and there is evidence that their initial concerns, about the erosion of the industrial base of the area, were important in changing the Corporation's original plans (*Interview* 5). From the outset of the petitioning process some members of the City Council had voiced their concerns over the land uses relating to St. Philips Marsh. There was concern in the area on the part of indigenous businesses that land values would spiral to such an extent that they would be unable to remain there. In addition, the relocation of business in the path of the Spine Road caused disquiet among the councillors, especially ward councillors who voiced concerns about planning blight in the path of the road. Environmental groups were also involved in the petitioning on environmental grounds. The Secretary of State granted permission for the spine road despite the objections.

St. Philips Marsh

This is where the current industrial uses are concentrated. The key plans for this area are the infrastructure proposals which access the area. The UDC have divided St. Philips Marsh into two planning areas north of the railway loop line which will retain current uses, and south of the loop line which will link up with Marsh Junction.

This is a ten acre site. Now a joint venture with Hanson PLC, which also owns the Imperial Tobacco sites in South Bristol. The proposals for this area have altered since the initial Vision for Bristol document. The plans then included a Science Park and medical technology developments which would give the area 'high value added manufacturing' inward investment (BDC 1989). Originally this was seen as the premier development site, significantly it will be split in two by the Spine Road. The deal negotiated with Hanson PLC apparently underpins the private sector contribution to the Spine Road project. The plans for this area involve relocating 27 businesses which employ 569 people and the additional 'regeneration strategy' involves the further relocation of 9 businesses and 253 jobs (*Estate Times* 14/6/91).

Although the CDA is the largest section of the UDA, it is now more important for its infrastructural projects than its development sites. The support for current uses in the area was incredibly strong across the usually divided City and County Council and this was the issue on which they both chose to fight the strategic employment implications of the Corporation's plan. Nevertheless it is interesting to note that more energy went into fighting the proposals for dealing with the other flagship site at Temple Meads Kingsley Village (BDC 1989). The original plans of the UDC to get rid of the 'noxious uses which are inconsistent with the area's potential and which in the Corporation's view could and should be relocated' failed, on the whole, foiled by a consortium of public opinion, the traditions of politicians who came from a manufacturing background and others more politically concerned with the implications of the decline of manufacturing employment in the area.

Temple Meads and Kingsley Village

This is a 90 acre site at the Western tip of the UDA and now its most valuable site. The plans incorporate Temple Meads station as a gateway to the City. Within the Temple Meads area the Corporation's plans include 100,000 square feet of office space, 110,000 square feet of shops and restaurants, 185,000 square feet for leisure, entertainment and culture, a 250 bedroom hotel and 90,000 square feet of hotelier units. Across the Floating Harbour in Kingsley Village there will be in the region of 450,000 square feet of new and refurbished office accommodation, 300,000 square feet of new and refurbished retail facilities, a 500 space car park and 385 new homes.

Both the City and County Councils objected to the plans and there was a controversial public enquiry about the plans. The plans first of all contravened the structure plan. The city's objections were unequivocal. They suggest in their evidence to the public enquiry that the area allocated was too 'extensive an area' to be treated comprehensively. More importantly, as far as the city was concerned, the office floorspace exceeded the structure and local plan allocation by ten times. Moreover, the retail development threatened the already existing Broadmead Centre. Crucially it was the office density which caused problems for the City (Avon County Council. Proof of Evidence. Public Enquiry 1991). The case which was made by the County Council hinged on the definition of the term regeneration, which involves using enhanced land value as part of a strategy of achieving 'gains for public purposes'. The Development Corporation won their case. Although the Bristol Development Corporation has won the case in both of the public enquiries, actual development remains the poorest in performance indicated outputs of all the Urban Development Corporations

Economic Development and Urban Policy in Post UDC Bristol

During the thirty year period 1962-1992 there was substantial restructuring in the Bristol travel to work area. The nature of this process, the demise of the city's traditional manufacturing industries, the increased emphasis upon the professional and financial services sector and the transfer of new light manufacturing industry to the city's periphery, manifested a particular political and economic response to the development plans of the Bristol UDC: hostility, incomprehension and indifference.

Bristol established a diverse economic base prior to a late industrialisation. This foundation has given advanced research and development industries a distinctive position in the post war and contemporary city. Having developed an early association with consumption related industry on the back of its mercantile adventure in slave and related cargoes, Bristol's economic base remained heavily consumption orientated (Lovering 1983).

During the pre and post war era the South West's aerospace industry facilitated the development of the city's technical infrastructure and skilled supply of labour. This in turn would help it expand in the North American market, securing further defence related orders and supporting ongoing research and development. Bristol avoided the worst of the 1973 economic crisis. The shake out of manufacturing industry which took place in many conurbations and that was precipitated by the crisis, did not hit the North West. In turn the public sector did not draw upon central government funds for urban policy initiatives until after the riots of 1981. Instead the South West, heavily dependent on MOD orders, and its superiority in the aerospace and electrical engineering industry, was relatively safe from serious unemployment until the late 1970s. Consequently it has been towards the intergovernmental networks which sustained the MOD connection that Bristol's coalition building has been directed.

Latterly its early foray into insurance, commercial and office development cushioned the city against the 1981 recession. However this relatively healthy situation disguised some fundamental restructuring in Bristol's economy. Any handle that Bristol had on manufacturing industry in previous eras, has all but disappeared now. Currently, the twenty major manufacturing industries in the city have their headquarters in London (Boddy et al 1986). During the seventies the presence of largely local food, drink and tobacco industries in particular, suffered shrinkage. For example, paper production and distribution had declined, tobacco virtually disappeared. There was a total job loss of 30% between 1971 and 1981, but as if to cement the areas commitment to service and professional industries, employment fell by only 19%. While production staff fell by 37%, salaried and managerial staff rose by 14% (Lovering 1983).

Manufacturing had peaked in 1965, and thereafter, as industrial floorspace declined, floorspace devoted to covered warehousing and office development increased. The uneasy relationship between the manufacturing and the service sector and the parallel friction between indigenous and external firms began to develop during the 1970s when the change in the identity of Bristol became apparent (*Venue* 10/6/88).

This tension was played out in the City Council, where councillors from manufacturing backgrounds in particular, heightened concern about the future and exhibited a fear about the commercial and professional emphasis in the regions future development. In 1973 the Labour council leader stated that the preoccupation with Bristol becoming a regional commercial centre 'had gone too far' (Boddy et al 1986). In 1967 the regional planning document, 'Region with a Future', advocated a laissez faire approach to industrial development. The decline of manufacturing in the 1970s and 1980s was matched by a growth in the prosperous financial and professional sector and a concentration of public utilities headquarters in Bristol.

There is a specific spatial profile to the expansion of service, professional and distribution employment in Bristol. The city centre shed a proportion of its manufacturing jobs, while the fringe, in the borough of North Avon gained new jobs as a result of the proximity to the motorway network. During the late 1960s advanced research and development facilities moved into the outer area and there were substantial inward investments in banking, finance and insurance. Sun Life, Phoenix, Clerical Medical, London Life, Natwest and Lloyds Banks arrived during 1973/74. A second wave of business and professional services came in the early 1980s. In the year ending 1989, BIF gained 21,000 employees. The growth in these areas promoted a huge growth in office development to the extent that there was over-supply in the area by 1981.

> 'the country's breeding ground where new types of industry are being developed, innovative forms of business accommodation are taking shape and new patterns of industrial relations are being established...when looking back at development in the western corridor we may recognise that the base is being laid for a new industrial revolution with comparable impact to the one that took place two hundred years ago' (Boddy et al. 1986).

The manufacturing residue from the previous era was concentrated in St Philips Marsh. Since the sixties there had been a consistent refusal by central government to fund the redevelopment of the docks. The dock in Bristol declined in the 19th century and the port activities moved to Avonmouth. The historic harbour is within the present UDA and is now underpinned by consumption activities.

But the activity since the 1960s around the historic harbour has allowed the coalescence of various amenity groups in the City. In fact, the harbour area developed amenities for new consumer industries in the early seventies and the groups involved in using these developed an early resistance to the City Council's plans for the Floating Harbour, now a key site within the UDA. The Council wanted to infill the docks for industrial manufacturing and warehousing purposes but were prevented by the actions of the Bristol City Docks group, saving the site for consumption uses (Bristol City Council 1988). Thus the amenity groups, which were to become particularly important in the early eighties, had prevented the City from pursuing traditional manufacturing uses, promoting other uses that were more consistent with the following era. The result was to promote the city centre dock area for consumption uses.

Bristol's economic base had been successively underpinned by diversification strategies dating back to the pre-industrial trading activities of the merchant core. Its late industrialisation facilitated an easy movement into advanced defence related industries in the mid twentieth century. These were concentrated on peripheral sites and benefited from the public provision of a modern transport infrastructure. The commercial development of the city, in part predicated upon its historic links, flourished as a result of the mix of infrastructure, labour supply and technological advance. Bristol has also been experiencing the shake-up of fordist structures, but interestingly, because of its earlier strategies of diversification and the relative lack of an extensive phase of accumulation, the movement into post-fordist regimes is advanced.

The Selection of Controlling Networks

There is no clear dominant urban class alliance in Bristol, but rather a series of significant fractions of capital in competition. It appears that the UDC has become allied to one of these fractions. Here the local public sector has been unable or unwilling to craft a local alliance. The lack of coherence has appeared to spur Bristol's economy into further bouts of competition. Until 1993 the way in which the public sector and private actors operated in Bristol in the post-UDC phase differed little from the period prior to designation, the public and private interest in the city remained at a distance and the public private partnership, the Bristol Initiative, remained isolated. There was no joint vision of the needs of the city. The plans of the Development Corporation remained embedded in the City centre interests of commercial capital.

The period of greatest political activity around the UDC, within the local authority at the City and County Council level was immediately after the announcement until a year after the Development Corporation had arrived. In this a period a number of things happened: the planning department, through the Chairman, united most of the opposition to the UDC using Labour Party networks and urban professionals who were eager to preserve the recent gains in anti-growth, pro-equity and environmental strategies. The political aspirations of this group married happily with the territorial concerns of Council planners. It was the latter which were most prominent in the County Council's motives in pressing for greater advantage once the House of Lords hearing had taken place and the UDC had begun to negotiate on the code of consultation. During the run up to the House of Lords Committee, democratic arguments had prevailed from the political group, in the very clear ideological battle which was being fought with local government though these were underpinned by territorial arguments about the necessity of including certain sites in the UDA. Ultimately it was these arguments which were to prevail in the hearing. Once the UDC had been established under new boundaries, both authorities concentrated on securing rights which would ensure access to the process of decision-making via as many group processes as possible. These groups and processes were underpinned by democratic considerations.

The UDC, however, did not give any ground on the issue of participation. The board was dominated by a network of actors whose main collective interests were commercial development, linked to traditional city centre commercial, property, financial and professional sectors. Their only cross-over with the public sector was joint membership of the University Council and an overlap with the Housing Corporation and neither of these interests appear as prominent in the strategies of the UDC.

The public sector in Bristol saw the UDC competing for any funds which might have been available for inner city development and the left wing councillors saw it as as limiting their democratic remit. On the right of the Council, which was dominated by manufacturing old guard there was a perceived threat to what was left of its power base within the city. Within the officer corps, similarly, the UDC threatened to constrain its room for manoeuvre. The remaining private sector in Bristol, fragmented as it was, saw the UDC as a competitor rather than as a useful ally. This sort of central government intervention was new in Bristol and seen as ill conceived. Advanced manufacturing and distribution concerns on the periphery of the city did not need a new high quality environment as Bristol already had this facility.

The defence related industry did not need central government to channel funds in this way; they were more concerned with central government's actions arising from the end of the cold war. Its channel had always been a direct one. Competition over uses for the city and consumption was already reorientating the city and what was happening here seemed peripheral to the development on the outer rim.

In Bristol the lack of any perceptible alliance and underpinning coherence and the split between the UDCs and the other agencies of government created a minimum of interaction between the agencies. Where this might have occurred such as in the area of training for instance, it does not because there is no strategic overview in the region, although increasingly this has begun to change The UDC in effect is in a competitive position in relation to other agencies in the city. In Bristol many amenity groups are excluded from the decision-making arena. There are dual processes operating. Bristol therefore operated in a competitive structure both under local authorities and UDCs. Although local authorities had attempted to impose a co-ordinated structure prior to the designation it was only partially successful and was completely abandoned on designation. Bristol also sought to exclude any dissenters or opposition forces from the negotiating arena.

Representation and Accountability

In Bristol it was imperative for forces opposing the UDC to use democratic mechanisms because no cross-class alliance negotiation over rules could take place other than through formal democratic processes. Maximum use of these avenues was made by the local authority prior to the designation. These included drawing on the local press, the elected and professional arenas within the local authorities, the arenas of political party politics, special interest groups and the House of Lords.

The power struggles took place when the local authority managed to briefly, on two occasions, unite a rainbow coalition on democratic interests, land speculation and the nature of regeneration. Small manufacturing interests aligned with environmentalists over developments in the central development area, the Spine Road in particular. But the relationship between significant land holders in the area and central government powers meant that their powers were limited. The later struggle over Temple Meads and Kingsley Village pitted the UDC against the City and County Councils which were attempting to maintain the viability of the existing city centre against developments closer to Temple Meads.

Bristol's public sector has a poor relationship with central government. Within the sphere of urban policy and urban regeneration, they were not reliant on central programmes, though they clearly relied on government policy vis a vis the defence industry. They were therefore uninvolved in the network of relationships that many other areas had developed in a variety of urban partnerships. The tussle between central and local state seemed very clear in Bristol's case. Bristol was lambasted as a 'loony left' authority by inner cities minister, David Trippier, and achieved notoriety as being the only council in

England to block the arrival of the UDC. Yet in the light of other experiences this was paradoxical as it was by no means the most politicised of the urban left authorities. The far left of Bristol was contrasted with the other Corporations that had been announced at the same time as Bristol. Andrew May accused the Minister of 'bullying' the council when Trippier stated at a private meeting that he would use 'every weapon in his political armoury against Bristol'. But it was not the local element of the argument that was pin-pointed; rather the politics of the council, and therefore the argument was constructed on an ideological level.

Dawn Primarolo MP had been publicly involved in the fight from the beginning of the case. She was clear that she saw the UDC as an attack upon 'local democracy and Bristol City Council...a move to bring about more control over Bristol by Whitehall'. Later she was to say,

> 'Bristol is very important politically – it's the jewel in the crown to control Bristol City Council and it is a testing ground for other policies such as HATS' and 'the government is proving an ideological point – that the private sector can do it' (*Evening Post* 9/12/87).

The far left of the political spectrum, the Socialist Workers Party and the Militant tendency within the Labour Party preferred direct action and demonstrated against the UDC in a series of publicised actions, for instance when property developers were invited to a meeting with William Waldegrave, organised by Hartnell Taylor Cook at the Grand Hotel in September 1988. The developers were jeered by the crowd outside while the city council sent officers in to 'spy' on the events.

It would not be true to say however that it was simply the nature of the policy process which created controversy in Bristol, the nature of regeneration policy became more important after designation. The row about the purpose of regeneration harks back to the pre-UDC era, when disputes took place in the city council about which economic strategy is desirable for regeneration. Central government actors placed an entirely different value on the economic activities which had taken place within the proposed UDA, as well as the type of social facilities that should be provided for the area. This was exemplified by Nicholas Ridley who countered Councillor Andrew May's concerns that the industrial reservoir of land available within the UDA would be sacrificed:

> 'They (the City Council) are trying to make this into a place where metal is bashed. That's why it hasn't been developed. This preoccupation with metal bashing may not be the right course. They have zoned for industry. Why does it have to be industry, why not offices or warehouses?' or again 'what this area needs is a whole new community'(*Evening Post* 9/12/87).

Leader of Bristol City Council, Graham Roberston, saw the key area of the proposed UDA as the St. Annes Board Mills site, which had been zoned by the Council for industrial use:

> "If housing is going to be built in that area I hope it will be of the price range that the average person can afford" (*Evening Post* 9/12/87).

The sentiments concerning the changing role of the centre of Bristol were also echoed by the Regional DOE Director, George Wedd, who stated that:

> 'Bristol was a growing area for businesses providing financial services, but they were restricted at the moment' (*Evening Post* 12/10/88).

The implication was that the lack of city centre sites was depriving the city of inward investment. Local Council officers supported members' stating that the 'industrial area needed a gentle massage rather than open heart surgery' (*Evening Post* 27/9/88). This was further backed by Dawn Primarolo, whose constituency is comprised of unskilled, semi skilled and skilled manual workers. Eighty per cent of the people who work in the St. Philips area live in Bristol (BDC 1992).

There were limited channels of power in Bristol. Prior to the Select Committee decision the local authority and the press were used, but as the decision-making environment closed on the arrival of the UDC and as there were political changes within the local authority where the focus moved away from the Urban left, opposition and criticism of the UDC remains in the hand of the media, which are interested in the poor outputs of the UDC. There does not seem to be a focus within the locality around the idea of a transitional strategy for the city. There is no labour market strategy at all, and it has been this problem which the public private sector partnership has raised. There is no alliance, no coherent vision. The lack of an alliance seems to allow for maximum experimentation. Within the national public sector context of Keynesian state strategy, Bristol's technological mix and local competition flourished, but it seems to be genuinely caught at a difficult period, with little capacity on the part of local organisations to manage the current situation.

The Policy Process

The decision framework in relation to the practices of regeneration within the UDC and on the part of the UDC does show a significant transition in line with the ideological motives and organisational imperatives of Central government.

There has been a move from strategic to flexible working arrangements, from departmental to team organisation and from in-house to external contracting of work. In addition there is competition between agencies and a

Figure 4.3 Decision-making Frameworks

Framework	Bristol Local Authority	Bristol Urban Development Corporation
Desired planning framework	Strategic	Non strategic, flexible
Desired internal organisation	Departmental (small)	Small teams
Desired personnel practice	In-house	Consultancies
Structure of competition	Competition between local authorities and also central government. Public private sector competition	Competition between all agencies public and private. Competition with central government
Structure of co-ordination and collaboration	Inter-departmental	Within contracts
Desired interagency interface	Non-hierarchical, limited strategic efforts	Fluid, ad hoc, hierarchical, non strategic

non-strategic organisational interface. Although this is consistent with the aspirations of ministers responsible for developing the initiative, there are two problems which can be observed.

Firstly, the movement from one paradigm to another is more or less limited to the UDC, though recently, City Challenge has shown some similar organisational characteristics, so it is questionable whether the UDC has had a hegemonic influence in the area. Secondly the framework which we observe in Bristol, seems as incoherent as it always was, and it is also questionable whether this framework could encourage the spur of a 'post fordist' accumulation regime. In Bristol then, the organisational framework seems to have developed as the government would have wished. There is clearly a move to a new type of framework. However we do see in Bristol that where there is more discretion i.e.

Figure 4.4 Decision-making Styles

Style	Bristol Local Authority	Bristol Urban Development Corporation
Dominant value	Public Good	Economic Good
Dominant ethic	Professionalism	Private sector professionalism
Dominant skills	Managerial	Opportunist and internal managerial
Attitude to competition	Intersectoral, inter-departmental, inter-organisational	Inter-organisational
Approach to development	Co-ordinated and planned	Speculative approach to development
Rules of engagement	Rules clear, participation and action clearly part of democratic framework	Rules ad hoc, ill determined, contingent on UDC
Legitimation of economic development planning	Legitimation of economic planning decisions through state form and democratic mandate	Legitimation not sought. Loosely based on internal ethics and performance of existing agencies
Approach to opposition	Co-option of some limited interests though unable to maintain stabilisation	Delegitimise and exclude oppositional interests from process
Approach to consultation	Informative, participative	Post hoc informative
Control of implementation mechanisms	Control of provision	Regulation of external groups
Desired Decision-making mode	Rational	Ad hoc

the structure or competition and collaboration, because there is a lack of a strong local/public sector influence the agency remains isolated.

The nature of the decision-making style shows a more complicated picture. It would appear that there has been the development of a substantive new style, this is especially clear when one examines the rules of engagement, the legitimation function and the approach to opposition and consultation. All of these would indicate minimal public interference in the economic development and regeneration field, and the development of new measurements of performance for public agencies operating in this field. Similarly there is also an expected movement in the control mechanism of the agency and of the overall approach to a mode of decision-making. But while the dominant value expresses a movement by the agents away from the Keynesian value system, the ethics and skill reflect the isolation and backwardness of the personnel involved in the UDC. These might be speculative and competitive, but this does not appear to be a result of their entrepreneurial ethos, rather they are hemmed in by the development opportunities which surround them. Again because of the lack of a public sector presence it seems that the introduction of 'new' ideas of agency development is easier.

Overall the UDC seems to have had little impact on the politics of Bristol except in so far as it has closed further avenues to urban funding which central government may have offered. But indeed it has had less impact on the processual direction of local state policies in Bristol. State strategies of distribution were as much about the advanced manufacturing sector as the Keynesian state processes of education, health care and housing.

In this sense the size of the UDC and its budget makes it a tiny component in the concentration of selective redistribution strategies. Similarly the existing labour market within the region and the lack of labour market activity on the part of the UDC suggest that its role in promoting the divisions between the polyvalent worker and the unskilled worker is negligible. Although non-democratic processes are on the increase, there has always been a dual system, only in the most recent decades has economic development been subject to economic control. Bristol seems marked by competition in organisational form in the public, quasi-public and the various private sectors. Within this competition the community have little or no capacity to affect any of the decisions. The evidence in Bristol is in the level of fragmentation that has meant that central government has had little impact on reformulating local plans or constituencies.

5 Trafford Park Development Corporation

Introduction

Trafford Park Development Corporation was established very quickly. The speedy designation was a result of the existence of an urban regeneration strategy that had been drawn up by the local authority, Trafford Metropolitan Council, and upon which the new Corporation's objectives would be based (Roger Tym and Partners 1986). The Trafford Park UDA is one mile to the west of Manchester City centre. It includes the existing Trafford Park Industrial Estate within the borough of Trafford, and the old British Steel site, Northbank Industrial Park at Irlam five miles away, a total area of 1,267 hectares. The problems confronting Trafford Park Metropolitan Borough Council in 1987 were technical rather than political: land assembly problems and outdated infrastructure, a lack of available finance to develop the area, competition with Salford Quays across the canal, and the redevelopment of Manchester city centre. At Irlam the main problem was environmental degradation as a result of British Steel's activities.

The Announcement and Designation

The local authority proposals for the UDA were put before the Secretary of State in 1986. Nicholas Ridley refused them and designated an ordinary Corporation, under the 1980 Act. The Trafford Park Development Corporation established primary objectives relating to the site and a series of secondary objectives covering employment and social regeneration (1).

The previous plans outlined in the investment strategy provided the tangible base for the UDCs policy and programme. The strategic objectives were also very similar. The important difference was in the administrative processes which were proposed and the specified relationship between the actors and their respective agencies. Trafford Borough's objectives were broadly met by the Development Corporation. The Regional Department of Environment envisaged few problems with the imposition of a central government model of a UDC, other than with the traditionally left-wing Manchester City Council. (*Interview* 7).

The model which the Borough had proposed was largely corporatist. The proposals for the working community mimicked a German model of involvement, whereby the workforce would be represented at board level in the proposed development agency, and the unions would prepare a joint statement of their willingness to discuss flexible manning and labour relations agreements to suit the technology and production methods of the firms with prospective investors.

The statement could be used in marketing the Park and furthermore unions would nominate a prospective investor who would negotiate with the firms. There was no residential community in the UDA thus obviating some of the more common community initiatives that were by this time emanating from LDDC.

The authority, and indeed the other agencies involved in the Trafford Park Industrial Strategy, wanted to actively restructure the local economy and promote it in a way that that they associated with the industrial estate at the height of its success:

> 'This was to be achieved by a single estate owner able to control the disposal and the initial use of the land to recover the proceeds of land disposal' (Roger Tym and Partners 1986).

Control of the land and, as a result of this, the development process, was seen as the most important component of the attempt to develop the estate's inward investment potential. Control was to be taken out of the hands of private and inefficient landlords and given to a coalition of local agents, ranged across the public and private sectors. The private sector was to be largely representative of industrial capital and the public sector leaders were to be from the borough of Trafford.

Politics in Trafford Park

Trafford Park is traditionally Conservative at both parliamentary and local government level, (though it has now fallen into Labour hands and it had a brief flirtation with Labour 1985-1986). It is a Conservative enclave however that is surrounded by labour authorities. Throughout the late 1970s and the 1980s, the borough council established its closest links however with its immediate business community setting up consultation procedures for the development of the Industrial Park. The local authority officers sought to involve the trades unions as well as the managers, of local companies. They wanted

> 'a unanimous voice over the changes that were felt to be necessary in the estate' (*Interview* 8).

Corporate planning was nevertheless important to a borough council that linked the inner city core to the more affluent suburbs, hence dealing with divergent populations. Strategic planning assumed more importance in the authority as central government resources became more scarce, as recession and restructuring threatened the balance of the borough activities, and as other authorities in the area became involved in highly visible competitive strategies continue.
A new management ethos grew: 'It is the intention, through a system of performance review, to further sharpen the focus of management to achieve service delivery in an efficient and business-like manner' (*Interview* 9).

In some senses Trafford followed a number of other authorities in responding to externally imposed conditions of expenditure restraint and cultural change.In fact much of the political heat came from the neighbouring Manchester City Council, while Trafford's position in the suburban green belt of Manchester made it one of the 'sleepier' hamlets of North West politics.

Urban Policy in Greater Manchester

TPMBC set up an economic development unit in 1979 and like many other local authorities there was immediately departmental rivalry between it and other departments (*Interview* 8). TMBC had a history of public and private sector co-operation that dated back to reorganisation in 1974. The Economic Development Unit originally had five members of staff – an Economic Development Officer and assistant, a small business officer, a tourist officer and administrative support. At the beginning of the Urban Development Corporation's life, three of these officers left with the Chief Executive to become members of staff at the Corporation (*Interview* 9). This departure, and the work of the Corporation, also signalled a reorganisation within the authority. EDU amalgamated with the Estates department and reorientated its work to deal with private sector enquiries into land availability and liaison with the businesses outside the UDA.

The Borough Council's earlier concern about the Industrial Park was flagged up after local government reorganisation in 1974 when the areas' firms had voiced concern about the deterioration of the Park's infrastructure. A consensus emerged between the authority and the business interests and in 1979 the Trafford Park Industrial Council was established. The Economic Development Unit of TPMBC approached the Managing Director of GEC with a request to lead this council. TRAFIC was an important spur to initial activity by all of the actors involved with Trafford Park. Managing Directors from seven of the leading companies were the driving force of TRAFIC. The trades unions were also part of this organisation. The firms included: Ciba, Geigy Industrial Chemicals, CPC UK Limited, GEC Limited, GKN Steel Stockyard Limited, ICI PLC, Kelloggs, and Procter and Gamble.

This attempt to galvanise local involvement met with intermittent success. One problem was the size of the park and the deep-rooted decline in the infrastructure. Additionally TPMBC had to juggle priorities with the rest of the Borough, and, as the 1980s progressed, it found increasing difficulty in getting hold of the capital resources to deal with the problems (*Interview* 8). A spur to further organisational action came in 1984, when the DOE announced that only £100,000 would be available for the borough in Assisted Area finance. Senior officers at the time were developing a series of plans for the development of local initiatives, formulated in consultation with local businesses, but found that they were short of finance. There was also, within the officer corps, a very deeply held frustration about the unwillingness of the borough to spend money on innovative projects for the park (*Interview* 8).

In the early 1980s, there was therefore a combination of problems which were chiefly identified by the officers of the local authority. The local authority needed a vehicle for change and resources for change. It had yet to agree upon the means to do this. A further difficulty was to identify and unite the various interests operating within the area and decide the type of modernisation that should take place.

Preceding Strategic Framework

At the beginning of the 1980s, there was a cumulative concern about the Greater Manchester economy in both the public and private sectors. Each of the local authorities in the central core undertook strategies to combat the problem. Salford developed a prestigious waterfront mixed development regeneration scheme. Manchester promoted a range of urban left strategies (Wainwright 1989). Later Manchester adopted the more traditional growth and property led strategies of the pragmatic metropolitan authorities. Trafford Council focused its efforts on the industrial park.

Until 1986 the planning framework was under the control of Manchester County Council. It has been suggested that local government reorganisation in 1974 and the creation of Greater Manchester County Council had a positive effect in engendering a strategic approach to the querulous economic and social effects of the post-1973 instability (Healey et al. 1988). Nevertheless, there are thirteen districts in the Greater Manchester conurbation, and the political differences between them remained acute in the early 1980s, especially about the issues of housing, renewing city centres and environmental reclamation. There was a concentration of economic development initiatives in the New Towns which meant that the County had

> 'to generate a conurbation-wide discourse about strategic matters in a context where the boroughs had grown accustomed to thinking independently' (Healey et al. 1988).

The County also had to mediate between the boroughs, which were fighting for political resources. And while there were market-led discrepancies between the areas, so there was also a difference in the resources they received from central government. Until the 1980s these two forces contrived to polarise the approaches and the strategies of the districts, Inner City Partnership status given to Salford and Manchester and Urban Programme status given to the other boroughs. There was therefore unusual room in the conurbation for district rivalries to provoke further fragmentation in public sector approaches to growth and decline. Not only did districts within the conurbation have different political traditions, they also experienced the discrepant effects of boom and recession, and their populations represented significantly different employment and social groupings (Law 1988).

Greater Manchester Structure Plan (1981 and its revision 1986) should have given a degree of coherence to the policies of all the participants. But there remained a tension between the districts who wanted to promote development and the County, which wanted stability and management of growth. Healey et al. suggest that local authority officers wanted expansion on peripheral greenfield sites, particularly in the south, towards the airport. But GMCC was concerned to stem the decline of the regional core and sought to balance this decline with the growth in the outer conurbations. Structure plans gradually lessened in effect as the eighties progressed. As strategic statements they might be useful but, as Healey and others have stated, the Counties lacked authority to deal with problems arising in the implementation of the plans (Healey et al. 1988). Soon after the demise of Greater Manchester County Council, each district was asked to prepare a Unitary Development Plan.

Aside from the Structure planning process, economic development was another feature of the metropolitan County Councils. They set up an economic development agency, Greater Manchester Economic Development Corporation. Certainly there has been a plethora of central government urban initiatives visited upon the area since 1977. Manchester and Salford, with the Old Trafford section of the borough of Trafford, were designated as an Urban Programme area in 1987.

Latterly new strategic guidance was issued for Greater Manchester by the DOE which gave the district Councils the basis for their Unitary Development Plans. The guidance encourages full public private sector co-operation, particularly in urban regeneration projects. It stresses the government's enabling function, pump priming grant aid in Manchester, and developing the international dimension of Manchester's work, and emphasises the expansion of the airport. In addition the Secretary of State continued the presumption against the release of peripheral open land unless it was shown to be consistent with urban regeneration. Manchester's town centre would continue to be the main focus for the provision of shopping facilities the regional shopping centre. Therefore in the late 1980s one could still see significant remnants of the Keynesian planning process in the region.

Trafford Metropolitan Borough and Urban Policy

In 1983/4, TPMBC, together with the local industrialists, set out its aspirations in a document entitled 'A Strategy for Revival and a Programme for Action'. The document recognised the challenge for Trafford Park implicit in the restructuring of the international economy and the consequences likely to be vested upon the borough as a result of global reorganisation of capital and labour strategies:

> 'The main aim is to lift Trafford Park out of its position as a decaying early 20th century industrial estate into a modern park equipped for the 21st century' (TMBC 1984).

The structural changes raised a question about the type of strategy which should be undertaken:

> 'It (Trafford Park) reflects the changing structure of industry in Britain and raises the question of whether or not it is realistic to expect the full extent of Trafford Park to be developed for industry again' (TPMBC 1984).

TPMBC felt that it would be impossible to look for comprehensive redevelopment of non-industrial uses, even though this was desirable. There was obviously debate in the authority about the merit of industrial versus non-industrial uses and the significance of industry in the park as a whole. The report discussed the amount of vacant industrial floorspace which by 1984 had risen to two million square feet:

> 'To some the present vacant floorspace and land is simply a trough in cyclical change and as the national economy prospers so will Trafford Park. To others it has long term significance. It reflects the changing structure of industry in Britain and it raises the question of whether or not it is realistic to expect the full extent of Trafford Park to be developed for industry again' (TPMBC).

In the end, it opted for 'a mixed industrial economy' which, meant there had to be some realignments in land ownership so that new uses and relocations could be zoned within the Park. This would: 'clearly involve the council in a much more interventionist role in determining the shape of future development than before' (TPMBC 1984).

TPMBC was not entirely happy with a simplified planning regime which they saw operating within the nearby Enterprise Zone. It felt it minimised its control over the quality of development that was taking place in the area. The strategy was highly interventionist and planned intervention was seen as crucial to its success:

> 'the implication of the broad philosophy outlined above is that together the industrialists, landowners, local authority and Central Government will act to steer both the direction and the pace of change in the park' (TPMBC 1984).

The difficulty between the landowners and the authority was implicit in the consideration of working roles as Trafford Council felt compelled to say that there had to be a willingness to work together and that the landowners had to accept a change in the standards of their caretaking of the land. From the industrialists it wanted a commitment to investment and to new product generation within the Park.

Other partnerships were also significant to the strategy. The Manchester Ship Canal Company was one of the partners, which together with Greater Manchester Council and Salford City Council, sought to do something about the threatened closure to the upper reaches of the Manchester Canal. Already Trafford was discussing with Greater Manchester PTE the idea of a light rapid transit system for the Greater Manchester area that would reach into the Park, as well as an attempt to improve bus routes.

TPMBC engaged in a political strategy which involved lobbying at central government level. In fact, once the bid had failed, later reports were to talk about a 'concerted campaign by firms, unions and public bodies interested in Trafford Park to achieve assisted area status' (Roger Tym and Partners 1986). Although they had attempted to secure designation for the area, within the Manchester and Salford Inner City Partnership, and managed to get assisted area status, they could not achieve any real impact within the Park.

The change that was envisaged to consolidate investment opportunities was typical of many local authority strategies developing at that time. The characteristics of implementation were to maximise the use of scarce resources, and the involvement of a whole range of agencies, i.e. environmental agencies such as Groundwork Trust, the European Commission, and the Manpower Services Commission, both because they were needed to attract finance, and because they would consolidate an area approach. They would target two or three locations at a time, the areas divided into the East, in which they wanted to place non-industrial uses (this became the Wharfside site), and the West, where they preferred the option of a technology Park, within UDC plans this became the Dumplington site. The Village continued with the same uses under the UDC as in earlier plans plus a range of highway and environmental improvements.

The proposals were officer driven. To some extent they were stimulated by the development of Salford Quays, across the canal in the City of Salford. The Quays promoted mixed development funded largely under Action for Cities and European finance. This facility had been developed by the Technical Services Directorate of Salford City Council and provided an exciting demonstration model for other local authority officers in the conurbation.

But without resources to implement the plans, the officers and the members of the borough were left with a stalemate, though, in 1985 the officers briefly felt optimistic because a Labour authority might spend more money on some of the development plans. The officers' response to the stalemate was to oversee a transition in TPMBC's policies from an orthodox managerial style approach to economic development towards a more speculative approach using pre-existing plans as a foundation.

It was obvious that the officers and members could not secure the changes in the park that were necessary as the gradual decline of resources available to local government for the autonomous implementation of programmes left them without finance. A number of problems confronted the policy-makers in Trafford Park. Much of the Park was not in public ownership. Both the authority and the major companies in the area had expressed dissatisfaction about the way that Trafford Park Estates PLC administered their landholdings (*Interview* 7). There was an abundance of vacant and underused sites in the area, marketing of the sites was poor, and there was a general unwillingness on the part of Trafford Park Estates PLC to do anything about marketing the sites, or indeed to explore the potential of the area.

The powers and resources available to TPMBC to correct this situation and the ability of firms to intervene in the context of traditional approaches seemed negligible. It seemed that the only way that local government could spearhead a regeneration strategy would be through an intermediate agency. Despite the Conservative majority in the authority, it appeared difficult for the members to assimilate the idea of an arm's length implementation company, even though this was a politically desirable strategy to central government. Trafford Park Investment Strategy was a joint venture between Trafford Council, the seven manufacturers who were involved in TRAFIC, and the two regional arms of central government. Roger Tym and Partners drew up a feasibility study for the implementation of a series of plans similar to those contained in the earlier Strategy for Action.

There were two possible types of strategy open to the authority – limited intervention or positive restructuring. Councillors opted for the latter. They wanted to create a special type of development corporation, emphasising partnership. This was especially because of the experience that the Chief Executive and his Commercial Director had experienced in visiting LDDC (*Interview* 9). They wanted a mechanism that would co-ordinate public and private investment. Their objectives were to attract other participants to prime the development process. This would involve leverage of £550 million worth of investment. A specific aim was job creation. In addition to construction jobs, 16,000 extra jobs were expected to be created (TPMBC 1986). Within the Trafford Park Industrial Strategy, Trafford Wharf envisaged the establishment of a regeneration agency committed to developing key flagship sites. Voluntary negotiation and relocation of existing businesses would be necessary to secure new uses. These uses would be consumption -oriented activities, high tech office uses, and leisure and recreation uses – a museum and pub for the dry dock area. A new bridge would be built connecting the area to Salford. A new dual carriageway and road would connect to the Pomona docks. Industrial development would occur to the south of the dual carriageway. Recreational uses would emerge at the Bridgewater canal (TPMBC 1986).

Essentially the plan was about not only restructuring the agency relationships of the area (the object of discontent here, being the intransigent local landowners), but also about restructuring the investment pattern in the labour market potential of the area. The plan focused on the type of agency that would be established (Roger Tym and Partners 1986). The institutional framework which evolved to support these developments involves very specific attempts by the public sector to hold control of the agency. Local accountability is maintained in the spending of public money through the involvement of democratically elected members.

A mixture of powers would, in fact, be distributed between the local authority and the agency. Vesting and voluntary land agreement lay with the agency, CPO responsibilities and highway powers and development control would lie with the authority. Grant allocation and distribution was to be administered through the agency.

Trafford Park Investment Strategy was politically unacceptable to central government in the climate of the late 1980s. Central government needed to control the allocation of public expenditure, maintain a commitment to the inner cities and be seen to be non-interventionist. In preparing the strategy, TPMBC had interrogated the dimensions of models like a UDC or an SDA, and had decided that they wanted an agency similar to a Development Corporation, but they also wanted to retain planning powers within the local authority. They were aware that they needed capital expenditure, and they were pragmatic enough to realise that they would only be able to obtain this finance by agreeing with central government's political objectives. This was exemplified at the time by the comments of Barry Brotherton, the Labour group leader of TPMBC, who said on the announcement of the UDC:

> 'We could have done all this ourselves if we had been given the money by the government. But what ever doubts we had, we knew we were not going to get the cash any other way. We couldn't say no' (*Independent* 5/5/1987).

The first Chairman of the Corporation said at the same time that he accepted that the UDC was in essence 'dictatorial' but was quite sure that they could have not developed an agency independently because they did not have access to finance.

The Urban Development Corporation

Establishing the Boards and Management Structure

There was considerable continuity between the key actors involved in the TPIS and the Urban Development Corporation. There was also a fair degree of horizontal networking between the actors of the UDC and those of other agencies in Greater Manchester. As noted above, the Chief Executive of the UDC had previously been the chief executive of TPMBC, and had been personally responsible for bringing the strategy into fruition. He took four of his local authority staff with him to the Development Corporation where he then recruited a predominantly public sector staff (*Interview* 8). This ensured considerable continuity between the practices of the UDC and those of surrounding public sector agencies. There was a feeling that they knew their way around the area and they quickly established a number of conurbation-wide officer fora (*Interview* 9). These networks were, however, officer based and it was emphasised on a number of occasions that the same relationship did not extend to the members of the local authorities (correspondence between TPDC and Manchester City Council).

The board of the Corporation included substantial public sector representation, the Conservative and Labour Leaders from TPMBC as well as the Deputy Leader of City Council. The leader and deputy leader of TPMBC and Salford City are both directors of Manchester Airport. A number of the directors have involvements with various public sector or quasi public sector bodies. The Leader of TPMBC is also a member of INWARD, the regional investment for the North West. One director from the private sector is a member of the North West Business Leadership Team and is Chairman of the Moss Side and Hulme Business Support Group. Another is the Director of the Central Manchester Hospital Trust. The Deputy Chairman sits on both the North West Industrial Development Agency and the Council of the University of Manchester Institute of Science and Technology and is Chairman of the Stockport District Health Authority. Five of the directors have some connection with the manufactures or the developers within the UDA. The Chairman was formally a Chairman of AMEC plc, which subsequently became the developers for the Wharfside Flagship. One director is a non-executive director of GEC plc, another is the Deputy Managing Director of the Kelloggs Company of Great Britain Limited and yet another is an Associate Director and Plant Manager for Proctor and Gamble. The final director is Chairman and Managing Director of Petros Group of Companies.Three of the directors have engineering backgrounds, one is a retired solicitor, one a surveyor, whilst the Leader of Trafford Council has a commercial background. Two of the original Board members have resigned since the incorporation of the Development Corporation. One of these members was a representative of a company involved in TRAFIC and a manufacturing company on the estate. The original Chairman of the Board, Mr Hadfield, carried with him numerous directorships, including the Chairmanship of the Greater Manchester Residuary Body and the Salford Phoenix. He was chairman of two property organisations as well as the Managing Director of Bass Limited and he held a regional directorship of Lloyd's Bank.

Unlike many of the other UDCs, therefore, there was no trade union representation and originally neither were there any female board members though this changed in later years. This, in part, reflects perhaps the homogeneity of the UDA, being almost exclusively comprised of company appointments. The absence of property interests perhaps suggests both the preferences of TRAFIC and the difficulties that TPMBC had with the property interests in the area, that is, they were seen as the problem. Labour interests appeared to be filtered through the community liaison group and, like the land interest, were seen as an interest to be negotiated with rather than to have their views directly represented.

Nevertheless there was a strong horizontal public private sector nexus developing in and around the board.

The executive framework for the Corporation was almost fully predetermined. The staffing ratio of the UDC is small and is spread across four directorates, Development, Operations, Commercial and Finance & Administration. However, the simplicity of this structure belies some of the complexities of the internal structure, which, as the Corporation developed, grew to resemble a standard large Economic Development Unit of a metropolitan council and one which housed a significant number of secondees. There were also a number of would be subcommittees, beginning in the second year of operation with a Community Liaison Committee. Later developments included a Trafford Park Performance and Quality Forum, and a Community Watch Committee. The consultative arrangements revolve around the Community Liaison Committee, chaired by the Labour Leader of Trafford Borough, Barry Brotherton, who is one of the council nominees on the board.

The Board of the Trafford Park Development Corporation meets monthly. In addition, it holds two monthly meetings to deal with planning matters. Although in its first year of operation it only dealt with 129 planning applications, the 1991/2 annual report states that it was through-putting 80 per cent of planning applications within eight weeks. From April 1991 the planning meetings were open to the public.

The Code of Consultation was established almost immediately with Salford City and Trafford Park Metropolitan Council, while an 'agreement' about consultation was reached with Manchester City Council. The Chief Executive argued that as Manchester City Council was outside its boundary they were not compelled to agree a consultation code, but, because of the transport proposals, it was useful for them to do so.

The administrative arrangement between Trafford Borough and the UDC is also different than that which was envisaged. One effect is that there is obviously a very different role for the Economic Development and Planning Units within the borough, though the Economic Development Officer comments that there isn't much difference between the work which the department undertakes currently and that which it undertook previously. But it is interesting too that his staff were not replaced and that the authority has been reorganised to deal with the needs of the Development Corporation rather than the other way around. They tend to have more responsibility for dealing with company displacement as a result of the relocations by the UDC (*Interview* 9).

Developing the Strategic Framework

There was a high degree of consensus about the strategic objectives of the UDC. This was the result of a number of factors. Firstly, it had taken four years to build a plan using a series of corporatist structures and involving the regional agencies, including the neighbouring Salford Council which also stood to gain from the development. Secondly, most of the proposals were around a very specific set of uses that were particular to this type of development, i.e. the type of development within an industrial park, and they complemented regional plans rather than competed with them. Where they did compete, as we shall see below, there were some problems. Unlike many of the other UDCs there was not a residential grouping in the area so there was none of the controversy that had been present in other areas where social versus other types of housing tenure was an issue.

TPDC undertook a joint programme with Salford City and Trafford Park Metropolitan Council and the Departments of Transport and Environment to improve the motorway network (2). A range of training in specific skill shortage areas and prevocational work carried out with the Manchester Training and Enterprise Council is emphasised within the work of the Corporation. This reflects the area's poor record in training and the preparation for new labour market strategies. This weakness has been identified by both the firms in the UDA and by inward investors and the training strategy forms a critical function in restructuring the local labour market. 138 environmental schemes are planned in the area, divided into highway corridors (which have included frontages to the main highways, as well as those of the major companies), gateways which include the Barton Road Swing Bridge, reclamation of eyesores throughout the park and public access to the canal sides, an ecology park, and refurbishment to community buildings, such as All Saints Church.

However, this is not to say that TPDC did not have a relationship with the trade unions within the estate. It has already been noted that a series of union agreements had been incorporated in the Trafford Park Investment Strategy, though these received scant attention in the later documentation of the Corporation. 'Social costs of UK labour supply' is a document given to inward investors by the Corporation. This contains 'half a page on labour relations' and states, 'Nowadays there is a growing tendency for terms and conditions to be company-specific and for workers and trade union structure to be more fragmented', but despite this the TUC chose to launch Trafford Park as the pilot for a new recruitment campaign. A preliminary research document prepared for the TUC had concluded that there was great scope for unionisation in the smaller firms on the state or in the UDA.

By 1990 the chief executive seemed to be worried about the lack of community involvement: 'we are trying to link more with the community around us'. Before this he stressed the existence of a 'community' within the park, a community which comprised the workforce of the various companies within the Park. In fact, not having a resident community seemed to be a positive asset in drawing

Figure 5.1 Trafford Development Corporation

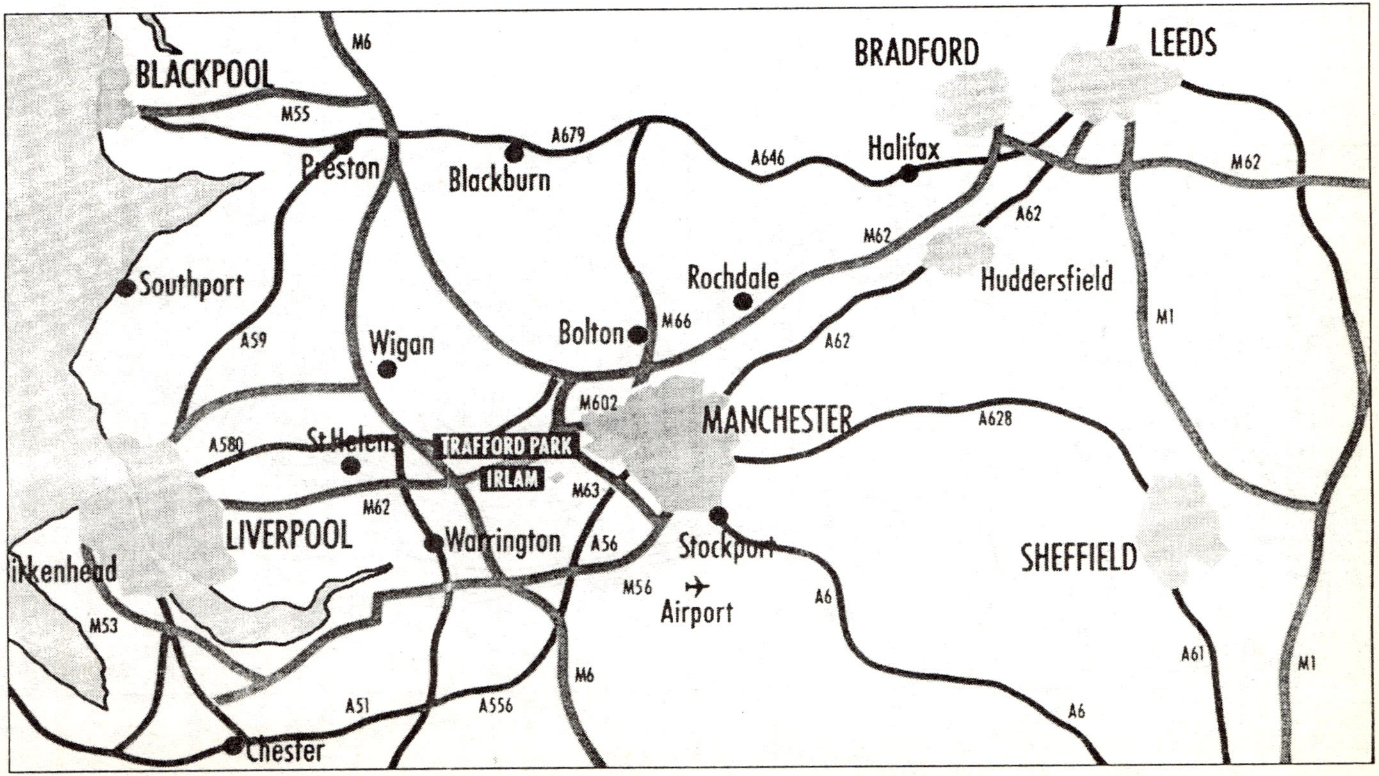

Figure 5.2 The Site

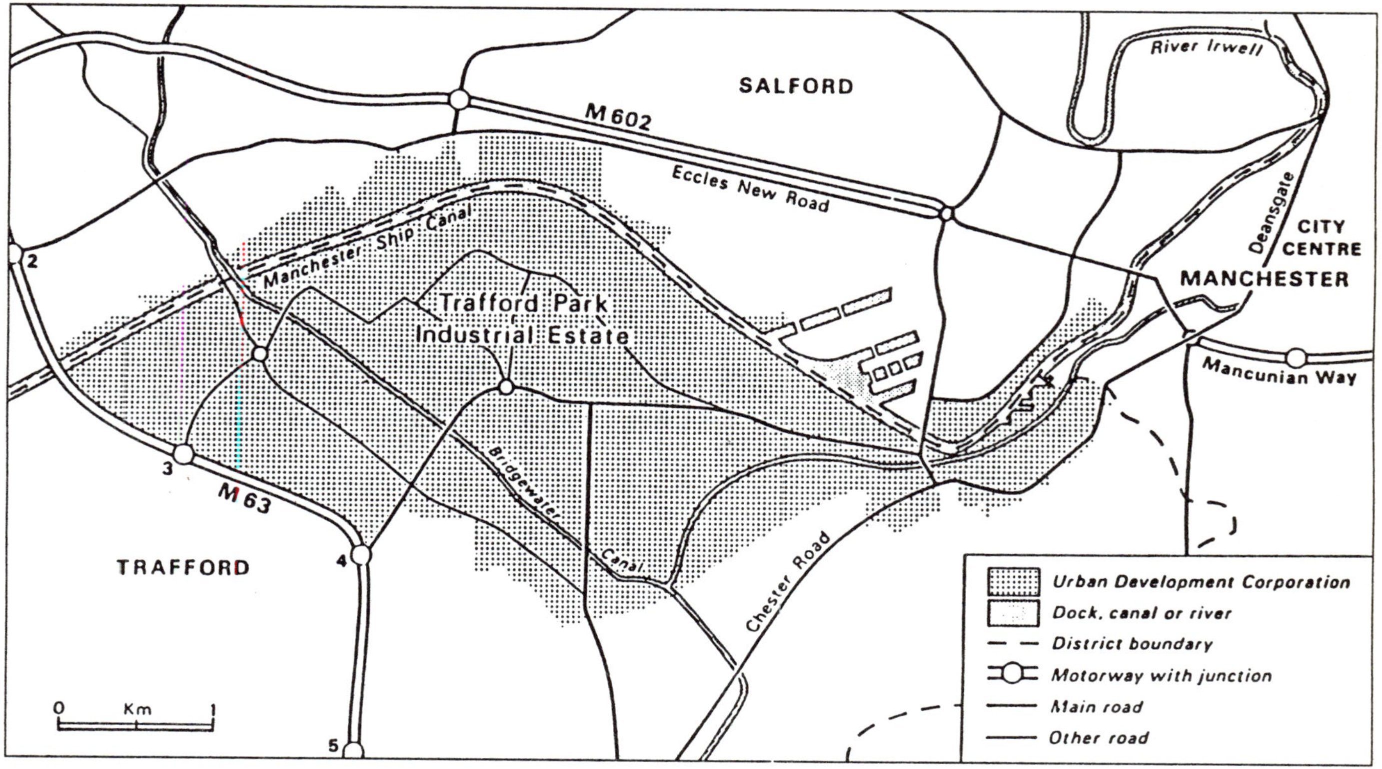

up the boundaries. Although they did not want to replicate the type of development that LDDC had undertaken, there were advantages in not having the type of problems that local residents had visited upon that area.

Other than this, the main links between Trafford Park Development Corporation and other agencies were officer networks built around functional tasks of scheme-orientated groups such as the North West Framework for Tourism Development. This latter group was established by Duncan Sutherland, of the English Tourist Board. He talks about their joint strategy with the Corporation as: 'A master plan to guide people in the right direction... We've been acting rather like a strategic authority'. (*Sunday Times* 16/4/89)

The overall decision-making environment was therefore extremely coherent within the UDCs boundary. The private and the public sectors espoused the same interests, but these interests were also congruent with the sub-regional strategy. Any individual conflicts with local authorities were often overridden by these considerations.

Inter-agency Relationships

The Council has formal and informal meetings with the UDC, using their nominees as the main point of contact. But the chief executive and his team meet on a monthly basis with the officers of the Corporation to discuss progress. There is also a progress report on a monthly basis to the Policy Committee of the Council. The Chief Executive of the Corporation is available to speak to any of the Council meetings that request his attendance.

TPMBC also has an agency agreement with the Development Corporation. The Planning Committee of the Council is informed of the Corporation's decisions, but they have no real power over the process, i.e. the local authority is consulted in the way that any other potential objector is consulted, diminishing its part in the entire development process. As the authority is the highway authority there are also day to day meetings between officers to discuss this element of their work.

However, the Urban Development Corporation is far more closely tied to central government and to Whitehall than had previously been envisaged, especially in its administrative processes, allocation of resources, and control over the programme via the corporate planning process. Its relationship with the borough is secured through three nominated board places, that are not technically accountable to the local authority, even though in Trafford Park they do, in fact, represent the boroughs of Trafford and Salford.

The Development Corporation has almost daily contact with the Regional Controller of the DOE (*Interview* 7). It usually concerns project direction. In the initial stages officers from the DOE were seconded to the Development

Corporation. They have a monthly formal meeting to discuss the large projects. The DOE do not discount disagreements between themselves and the UDC but declined to comment on the specifics of this, saying that they are usually over big schemes. As far as the Corporate Plan is concerned, the regional DOE and the UDC seem to be in general agreement. The DOE view the equanimity with which they deal with each other as being a result of lack of population in the area, though it was clear that its view was that in some ways it would have been better to have an area that was devoid of industry as well. As the officer in charge of planning said, strong negotiating skills were needed to sort out the firms in the area. The DOE sees its role as pushing the UDC to prioritise its activities. The main point of contact between the UDC and the DOE is over finance. There is a monthly claim covering all budget headings, from staff costs to loans, grants and acquisitions. The UDC has delegated powers only up to £100,000. In fact, the UDC deals with many of the same issues as the local authority, but aside from its relationship with the DOE, it has more freedom to manoeuvre in its development decisions. It has chosen to mimic the policy style of the local authority at officer level but has been able to extend the interagency contact which it did not benefit from prior to the establishment of the UDC. It must be said that the number of executive agencies operating in the Greater Manchester area is now much greater than prior to the UDC.

The Urban Development Area

The Manchester Ship Canal opened in 1894. Extending as far as the Mersey estuary, it was created in order to force Liverpool to reduce its charges for the importation of raw cotton for the Lancashire textile industry. In the early and middle twentieth century, the canal prospered. At its peak employment reached 3,500 but dipped to a low of 3,000 by the late sixties (Law 1988). Eventually it could not escape the ineluctable fate that had confronted other inland British waterways and ports. Road and rail transport were the inevitable beneficiary of the post war trade in transportable goods and publicly funded infrastructure. The programmes of the Keynesian new deal grew at a rapid pace. Containerisation added to the canal's problems as did changing preferences in port usage. The decreasing canal trade had a direct impact on blue collar and commercial and manual jobs in the areas of Trafford and Salford. Most of the commercial development at that time was based in the centre of Manchester.

By the middle of the 1970s very little trade was reaching the Docks and its owners, the Manchester Ship Canal Company, began to consider other uses for them. Between 1981 and 1983 MSCC sold the majority of Salford Docks to Salford City Council (Salford City Council 1988) but retained the Pomona docks, which had become part of the Enterprise Zone and eventually part of the UDA. By 1986 the North West Water Authority had precipitated a crisis in the company by threatening to withdraw its sludge boats which would close the upper reaches of the canal to commercial trade. This came soon after a takeover bid by the North-West property entrepreneur, John Whittaker. Soon after, a tripartite Ship Canal Steering Committee was established and the following year a ship canal forum set up comprising the public sector, trade unions, voluntary groups and an arm of central government (*Interview* 10).

Feasibility studies were undertaken by the MSCC and a modernisation plan was developed. The local authorities, Trafford MBC and Salford City Council, provided a short term subsidy for the canal while European finance was sought. Subsequently the Company benefited from a multi-agency funding package.

During its early years the fortunes of the canal were tied to those of Trafford Park Industrial Estate. This was the first such facility to be built in the UK and was constructed in 1896. It was connected to Salford Docks, across the canal, and linked to Trafford Wharf by its own road and rail company. The Estate became a magnet for Fordist manufacturing, distribution and wholesaling companies. Henry Ford located a factory for Model T Ford cars in 1911. Westinghouse (later part of GEC) built a model village for 600 workers there to accompany its huge factory in 1901 (Trafford Park Development Corporation 1989).

The peak in employment on the Estate (up to 75,000 jobs) came during World War Two. Post-war decline was intermittent, but by the 1980s, the Park experienced a multiplicity of problems. Trafford Park Industrial Estate PLC, which owned the estate caused problems for both the public sector and the firms which were based there. It refused to modernise the infrastructure: 'the lack of an organisation managing the estate as a whole meant that there were no means of achieving the orderly redevelopment of sites or the progressive upgrading of the infrastructure and environment' (*Interview* 9).

The industrial structure of the estate was outdated (many of the firms were operating in declining markets with no product development or research), and the lack of investment in modern plant and equipment, its poor image, and limited financial incentives made it an uncompetitive location.

While most of the site was derelict, at the eastern tip of the park an Enterprise Zone had been designated in 1981. The Enterprise Zone was 80% full by 1990; until then it was felt that the zone had contributed to the decline in property values elsewhere in the park However, by then there was little room for developing large prestigious schemes within the area, so the advantages of the rest of the park within the Development Corporation's remit became much more obvious. Within the Trafford part of the Zone, 3,961 jobs had been created. 1,480 were new jobs, the remainder were relocations. 740 were manufacturing jobs, the rest were service and distribution (Law 1988).

Trafford Park was in an unusual economic position. It felt the effects of decline and restructuring within Fordist industries but at the same time benefited from the potential which Keynesian state infrastructures created for the future development of markets. Additionally, although it was extremely close to Manchester city centre, it had not experienced the same level of decay, neither was it subject to the competing demand of a city centre space with little development potential. At the time of its designation there were 24,468 people employed across 562 companies in the area. The companies included Kelloggs, Ciba Geigy, GEC, Massey Ferguson and Courtaulds (TPMBC 1986).

While the UDA faced significant problems, it did not exist in an area of unmitigated decline. The structure of employment, however, had undergone a substantial transformation since 1971; 40% of the employment is in 12% of the firms in food, drink, tobacco, chemicals and mechanical and electrical engineering. A further 27% of employment is found in 52% of firms in transport and distribution. Recent gains, 1983-6, had been in transport and distribution; losses had been in manufacturing. Forty eight per cent of the firms in Trafford Park are branch plants; only five per cent of these are headquarters of a multi-plant firm. Only twenty per cent of branch plants had headquarters located in Greater Manchester and fewer than ten per cent have HQs outside the UK (TPMBC 1986). At the time of designation the employment level had dropped from a post-war peak of 75,000.

Like many other districts in urban conurbations, the area had been sensitive to manufacturing decline, though because of its strategic position to the south of Manchester, close to the airport and surrounded by a good motorway system, it retained significant modern uses and a good selection of factories engaged in modern labour processes. The Enterprise Zone had, of course, prevented the total decline of the area when other parts of the Greater Manchester conurbation were experiencing a more rapid decline.

The Village

The Village is a 55 acre site in the centre of the site and is the old grid development which Westinghouse (later GEC) built for its work-force in 1911. The aim of the Corporation in this area has been to improve the environment for small business by creating managed work space, office and conference facilities, small industrial units and studio accommodation. In addition the 'Village' seeks to provide leisure facilities for the work force of the entire estate. Like the other major development sites, the Village has been the subject of a CPO which was initially made in 1989 and went to public enquiry in January 1991. The Corporation owned 75% of the site by the end of 1992, and in an attempt to deal with the lack of an order for the village as well as difficulties in naming a developer, brought forward £5 million of highway developments providing improved internal roads within the village and better access to the site. (TPDC Annual Report 1992). By 1995 there were five completed developments in the village and four new projects were on stream.

Hadfield Street

This site covers 30 acres on the eastern end of the boundary, near Manchester City Centre. It was declared an Industrial Improvement Area in 1988, and 112 small businesses already exist there. The emphasis in this development has been on environmental and infrastructural work to secure the future for existing businesses and provide the infrastructure for the further development of small businesses. The CPO for the road network was made in May 1990 and confirmed in December 1991. Reconstruction of six internal highways has been completed. A 50,000 square feet business centre has been constructed by Laing Urban Developments using City Grant as well as a further 50,000 square feet of refurbishment of industrial and terraced properties in the area. City Link will run along the western boundary of the area while access on the A56 road to Manchester has been improved with an emphasis on improvement to the area as a Gateway. The emphasis here has been on interagency working as a result of the interface with Manchester and the importance of transport links (TPDC Annual Report 1993-4)

Northbank Industrial Estate Irlam

This 250 acre site was left derelict after the departure of British Steel in the mid 1970s. Most of the work here has been land reclamation, environment and infrastructural improvements. The highway improvements that have been undertaken jointly with Salford City Council have included the Irlam-Cadishead Bypass which improves access to the M6 and M63 and have also involved the upgrading of four internal roads. There has also been the construction of a port facility in 1990-91 by the Manchester Ship Canal Company which can berth ships of up to 10,000 tons. The site is a modern industrial park that has acted as a site for some of the business relocated out of the main Trafford Park area (TPDC).

Wharfside

Wharfside is an example of development in the UDA which can be used to look at the changes, conflicts and cleavages which have built up around the Corporation's work. It is deliberately chosen as the development which the Corporation has paid most attention to, aside from the Dumplington site.

This example illustrates the development process and offer an impression of the pattern of social relations surrounding the Development Corporation.

Wharfside is a 200 acre site at the eastern end of the UDA, between the Manchester Ship Canal and the Bridgewater Canal. The aspirations for Wharfside gradually changed over the period of the Development Corporation's control of the site. Originally the plans for the site were to construct:

1.8 million square feet of accommodation in total.
1.1 million square feet of light industrial and leisure uses. 500,000 square feet of offices (TPDC 1991).

In July 1990, the developers, Rosehaugh, in partnership with the Manchester Ship Canal Company, came forward with planning application for a much larger project:

	Square Feet	
Offices	2,137,250	(44%)
B1	418,000	(13%)
Industry	177,000	(5%)
Retail	75,000	(2%)
Leisure	120,000	(4%)
Residential	254,000	(8%)
Hotel	130,000	(4%)
Total floorspace	3,311,650	

The transformation in the objectives over Wharfside was predicated on a number of factors. The area's EZ status would terminate in 1991, and the attraction of development to the site prior to that date was extremely important. Rental value had risen in the area mirroring the property speculation boom taking place elsewhere in the county. Rents in Manchester had risen from £7.50 in 1987 to £9.50 in 1989 and were set to rise even further at that time (*Estate Times* 27/7/90). The new B1 use classes order came into being, allowing greater flexibility for property developers in developing high class office uses, high-tech and light industry in one band.

The mix of the agencies that operate in the field of planning and economic development in Greater Manchester means that neighbouring authorities are greatly concerned about each other's strategies. Manchester City Council saw the Wharfside plans as a threat to its own strategic policy proposals. In Manchester another concern existed. The members were excluded from the dialogue on Wharfside but within the authority there was a political commitment to such dialogue in the arena of agency relationships. In addition, Manchester City Council did not have a member involved in the debate at all in the area (Graham Stringer, the council leader, was, in fact, a member of Central

Manchester Development Corporation). Manchester City Council wanted to pursue an involvement with Trafford Park on its own terms and to ensure that strategic considerations were placed on the agenda: 'Both formal and informal machinery would ensure that the processes of implementing such a strategy through a series of complementary rather than competitive actions' (Manchester City Council 1989). The City wanted an agreement on the nature of development in the UDA, but was also keen to wrest out of the UDC a series of consultative arrangements using existing machinery or developing new mechanisms as appropriate. In the case of the city council, this extended to the proposal of a joint understanding on capital submissions to central government between the agencies in the region.

The proposals for Wharfside that emerged from the consortium of Rosehaugh/MSCC threw together a strange configuration of old and new agencies and a tension emerged in the private sector within the UDA and the old industrial estate as a result of the expansion of office development. Manchester City Council was worried about the scale of office accommodation suggested by the Rosehaugh Development, fearing that it would impinge on its own attempts to deal with increasing the potential for higher rates of office occupation in the city centre (*Estate Times* 10/4/1991).

The dialogue between the TPDC and the other bodies in the area over Wharfside was slow to emerge, appearing sporadic, patchy and inconclusive. Manchester wanted to negotiate the objectives and consultation procedures over the Wharfside development:

> 'Those opportunities (successful regeneration) will, however, only be exploited to the full if regeneration maximises the benefits to those residents in Trafford, Salford and Manchester in greatest need; is sustainable in the longer term; ensures sensitive and sensible development activity which is consistent with the pre-existing strategic planning framework supporting the regional centre; and results in net jobs over all'. 'These key requirements will only be met by common acceptance of objectives and awareness of how each of the organisations involved are tackling them'... 'I realise that this is a matter which has been raised before in the context of the consultation code but the City remains concerned that the code is still silent on strategic objectives and their interrelationship with other organisations. These are not capable of being handled effectively by inter organisation meetings on ad hoc issues' (Correspondence between Manchester City Council and TPDC 1988).

The City Council goes on to voice concerns about two schemes. The first concerns the LRT route and the traffic measures relating to the Chester Road roundabout. On Wharfside its interests became even more clear;

'The structural and functional relationships between the area and the regional centre are not explained. As a result we are far from clear how far, if at all, the study has assessed the viability of the range of new products or developments in the context of the regional centre. More particularly the apparent characteristics of these new developments are that they will duplicate those already found or planned in the regional centre. The references to leisure, shopping, offices, conference hall – even the relocation of the Halle – are cases in point....there is concern that the effect of the proposals will do little more than make substantial competition with the existing role of the regional centre, making a net increase in jobs difficult to achieve, quite apart from undermining the existing planning framework' (Correspondence between Manchester City Council and TPDC 1988).

This communication from Manchester began a flurry of letters and a series of meetings at the city council which played around with the decision. The initial response from the Development Corporation was tart:

'We are not in the business of stealing jobs from elsewhere in the county. Our targets are further afield and generally a lot further south' (TPDC 1988).

As far as the strategic objectives and consultation mechanisms involving members were concerned, both of the issues raised were seen as already adopted as 'policy' and 'non negotiable'. The Corporation's Chief Executive flags up the Greater Manchester Planning Officers Group as the 'prime forum for discussion'. Any involvement or joint meeting between the members of the authority and the members of the board will need to be 'extraordinary' because the pressure of time is an inhibiting factor. Replying to the specific issue on Wharfside, the chief executive uses the new B1 class as a way out of a potentially difficult situation of competition. For instance, he says that of:

'1.8 million square feet of new floorspace, 1.1 million will be light industrial B1 class. He stresses that the 500 square feet of offices will be phased, and he suggests that there will be little new office space available for over-demand in the city centre. An estate time survey confirms this. He also stresses that the out of town location, in any case, has a different market and that the shopping will be festival shopping' (Correspondence between Manchester City Council and TPDC 1988).

Manchester City Council, had its own regeneration strategy document from which its policy objectives could flow – 'The Heart of the Conurbation' (Manchester City Council 86). The 'Heart of the Conurbation' included Wharfside within its strategic ambit. The uses assigned to the Wharfside development were within this document. High quality landscaping, low density research and development facilities, a small amount of office space and the improvement of infrastructure and water courses, were all seen as complementary to the uses developed in the city centre as a regional capital.

Manchester had re-evaluated its city centre strategy by 1986. It wanted to keep office vacancy rates low. It continued the policy of maintaining a hierarchy of shopping areas in the conurbation, but it was also charged with maintaining a strategic view for Manchester. This was implicit in the joint relationships it tried to establish all over the region. This meant, for instance, that it had developed a very close interest with the Central Manchester Development Corporation.

In Manchester, the new urban left was fighting with the new managerialist or entrepreneurialist tradition in trying to resolve the conflicts. This is why we see a two pronged approach to the Development Corporation, on the one hand a straight-forward battle on the territorial division of space, an entrepreneurial competition over space, not only between regions but within them, with less mediation by the forces of accountability, and on the other hand a pressure to force some mediation into the process within a political agenda of consultation. Following the application from Rosehaugh/MSCC for the Wharfside site, MCC objected to the development and sought a meeting with the Corporation to discuss their plans. The City Council wanted to develop a common approach but were worried by the loss of land to non-industrial uses They were also alarmed by the proposals to develop a large out of town shopping site at Dumplington at the western tip of the UDA.

The option was for those interests who objected to the application to request the Secretary of State to call in the application for Wharfside. In fact, it was not as straight-forward as this, as the various agencies disagreed over what was necessary. The Council co-opted the Central Manchester Development Corporation to their case. Central Manchester Development Corporation was another agency who raised an objection in consultation with Manchester City Council. CMDC had been established as part of the third phase of Development Corporations. They had a small boundary and limited funding – £20 million pounds over five years – which made them quite reliant on the other agencies in the area. They agreed to commission a joint study of office and retail needs with Manchester City Council and they objected to Wharfside on impact grounds, traffic and infrastructure.

Rank Hovis McDougal were directly affected by the proposals, also expressed concern at the processes inherent in the UDCs decision-making and were concerned about the city council's capitulation pointed out that:

> 'TPDC does not own all the land involved in the outline application area and hence cannot guarantee any phasing programme at this stage. Equally once the outline approval is granted, the master plan objectives will be

> implemented and this does not include any policy for phasing'. (Correspondence CMDC and Lawson Price 1989)

They go on to to point out that although TPDC:

> 'may verbally agree to such an approach, there is no guarantee that the applicants, that is Manchester Ship Canal and Rosehaugh Trafford, will be equally agreeable to such an approach'.

They also suggest that it is impossible to monitor over provision because it could only be done once the damage has been done to the city centre. They continued to press for a DOE call-in, making charges of secrecy in the process. By 1990 the position had been eased, to a certain extent, by private negotiation. Manchester City Council officers met with Trafford Park Development Corporation and agreed a compromise, phased the development.

The Development Corporation agreed to control the development by managing the land as it passed through the development stages, giving TPDC stronger powers while land remained in its ownership, using a legal agreement and releasing land in stages while monitoring the impact of each of the phases. The Development Corporation agreed that they would be flexible about future uses to accord more with the council's Heart of the Conurbation Strategy. The corporation had commissioned a survey on Wharfside's impact on local offices which was tabled at the meeting, and agreed that it would be major headquarters rather than small suite lettings.

The City Planning Officer recommended at a subsequent meeting of the Planning Committee, that there had been 'a substantial change in the approach that they (the UDC) are attempting to adopt in relation to Wharfside'. Doubts remained about the overlap in the office market. But he argued that they had secured huge concessions from the UDC, over phasing, monitoring and flexibility. If this could all be negotiated, it would not be appropriate to persist with a formal objection and a request to the DOE for a call in. He asked for delegated authority to deal with the situation if Trafford Park did not agree with the conditions in consultation with the chair and the vice chair. So this is how Manchester resolved the Wharfside situation, but there were a number of other agencies involved in the area.

By 1990 this was seen to be a satisfactory outcome to the situation. Market conditions had completely altered. Although the Development Corporation had no statutory responsibility to acknowledge other agencies' local regeneration strategies, they stated their intention to do so, though it must be said that much of the agreement between all the parties was handled informally. Nevertheless, a crisis had been averted and the preferred developer was dropped. The first phase of Wharfside was complete by the end of 1992. Wharfside has gone on to be the development site for a commercial part and has attracted a derelict land grant. (TPUDC 1994-1995).

Economic Development in Manchester

Greater Manchester Conurbation is comprised of a cluster of west Lancashire towns and cities – Salford and Manchester, Bolton, Oldham, Stockport and Trafford, Bury and Rochdale. The population of the conurbation in the 1981 census was about six million (Manchester City Council 1992). The ramifications of the 1973 crisis were significant in the North West, consolidating debilitating post-war trends. Up to 1981, 90% of the job losses in Manchester were in manufacturing, where employment fell by 51%; it lost manufacturing jobs faster than any other UK region (Chestertons 1993).

By the mid-eighties, Manchester was a service economy. Seventy seven per cent of the jobs in 1985 were in the service sector. But overall the service sector was not growing. Since 1981 job losses in the service sector has been greater than job losses in the manufacturing sector. Production industries within the conurbation lagged behind the rest of the North West region and the UK in general, although the transport and distribution sectors, property, financial and professional services grew in the region (Manchester City Council 1987).

Within the Trafford Park Estate, similar growth occurred in chemicals, electricals, and electronic engineering. None of these industries had benefited from the overall pattern of growth established elsewhere in the UK. A spatial transformation was taking place alongside the restructuring of local labour markets. In the post war period some of the population of central Manchester emigrated to the new and southern towns of Macclesfield, Trafford, and Stockport (Law 1989) This was echoed by a similar trend in employment, which increasingly worried the core authorities.

The changes in the economy have also had an effect on Manchester's commercial office market. This has been affected by the decline in the key textile industries, concentration of business into larger units, facilitating the use of internal markets and the concentration of company headquarters in London. But additionally office development has also moved to the peripheral zones of the conurbation. Although during the 1960s and 1970s, there was a growth in office building in central Manchester, peaking in 1973, from 1974 to 1983 the supply of offices in the City of Manchester declined from 40% to 33%. Since then, there has been a decline in new office development accompanied by an increase in the refurbishment of offices in the inner area. By the mid eighties there was also a fairly high vacancy rate (21.8% of the total stock) (Law 1986).

By the late eighties the office market in Manchester was subject to fairly wild fluctuations in supply and demand, with under- and then over-supply following in quick succession, the result of both property speculation and the public policy proposals of the various public and quasi-public sector firms in the area. Greater Manchester experienced a number of contrasting trends during the 1970s and 1980s. Inner city decline was caused by the emigration of indigenous manufacturing industry, as well as the departure of regional headquarters and the financial services sector to London. Simultaneously there was growth on the periphery underpinned by the development of publicly funded infrastructure. The difference between the disintegrating core and the developing periphery is much more stark in Trafford Park than in the other case study areas. This is because the UDA straddles the inner and outer areas. The outer area has benefited form national infrastructure investment programmes, while the inner areas, in serious decline, have attracted urban aid. Therefore there has been no real continuity in the style of decision-making around these issues and the institutional fragmentation inherent in the environment of the 1980s did not substantially alter the problems which the UDC had to face.

In the North West there is a coexistence of intensive and extensive patterns of accumulation, though because of the nature of Trafford Park – an industrial estate built to house the factories of Westinghose and Henry Ford – intensive patterns of accumulation predominate. It was these industries that were decimated in the crisis of the 1970's and 1980's. Here too there is a move towards the characteristics described by postfordists. What is interesting here is that the focus is two fold. On the one hand local groupings which have developed in the UDA and on the other a conurbation wide new elite of officers developing strategic considerations. It appears that there is a high level of coherence existed at the point of transition to fordism overlapping somewhat with keynesian state management. But there have been discontinuities in the alliance. Recently it would appear that there is some movement towards building a new alliance. This is even more important to the locality, but difficult because of the mix of economic trends contained therein. Here they did not rely as much as Tyne and Wear did in the post war era on regional policy, rather they depended on national keyensian state infrastructure projects. Trafford also stood to benefit from the redistributive policies targeted at Manchester City Centre which were seen as complimentary to those of Trafford. Such a confluence of local economic scenarios in Manchester made it difficult to achieve political coherence during the Fordist period. Trafford does not therefore experience the continuity in a ruling class alliance, nor does it have to experience the contrast which this would impose on their technological development. Although however Trafford is subject to many changes in organisational form, it may be hampered in pursuing them by the domination of traditional officer skills from Trafford borough.

Although the central state is far more active in the day to day decision-making in Greater Manchester than previously, not only through Trafford Park Development Corporation but also through the remits of Central Manchester DC and the Training and Enterprise Council, a network of new agencies has sought to develop strategies to maximise the sub-regional distribution of those resources. This involves significant inter-agency co-operation at officer level. Within the local authority, back-bench politicians have very little to do with this process. Leading councillors are far more likely to be involved in this arena, at the very least through their membership of the new agencies.

The area is characterised by fragmentation and a lack of coherence in the previous era, leading to fragmentation in any periodic ruling alliances that might exist. This meant that internal regional competition between institutions and sectors was substantial. As declining resources in redistributional funds from central government became marked, a coalition of professional officers working across agencies developed an inter-agency network. This was easier to do in areas like Trafford Park where there was no indigenous population and much more difficult in areas where there had been active participation, like inner city Manchester. However, the inter-agency networks are effectively the decision-making arena for Greater Manchester.

Within the UDA there is a consensus between the Development Corporation and the constituent companies about the strategies which have been employed and about the processes which have developed to implement them. This involves a sophisticated labour market strategy as well as development-led regeneration. In Trafford the consensus about policy form was broken by a neighbouring authority working in consultation with the UDC and other interests in the park. The break however was over a strategic issue. Manchester City Council were entitled to object to the UDC plans for the provision of office space at Wharfside. In doing so however they sought to incorporate demands for member representation in UDC Council negotiations. This was discretionary and was refused. Trafford in the context of an emerging alliance of officer and professional and a weak property market gave way on a strategic issue but held ground on a procedural one.

Negotiating in this environment was something which the UDC officers were willing to engage with but were happier to do so without member 'interference'. This in itself reveals a conflict, which is clearly played out in the Wharfside case, as Manchester City Council is one of the few agencies charged with regeneration in the area which still operates through a committee structure. This, and Manchester's residual commitment to equal opportunities and equity strategies, meant that the most likely agency to experience difficulties with the Corporation.

However, a regional coherence around new consumption uses aimed at facilitating inward investment grew in the area. Within the technological mix transport infrastructure would become crucial in galvanising all the public authorities and quasi-state agencies. Most of this development was in consumption and distribution related industry, and therefore the UDA had a significant role to play as far as the region was concerned in providing space for advanced manufacturing. This in some sense was challenged by the Wharfside and later the Dumplington developments, as the agencies had already established a *de facto* division of labour market structure and developmental strategies. The row between Manchester City Council and the Development Corporation was as much to do with these territorial concerns as it was about the political demands associated with the involvement of members. The consensus around the strategies of the UDC was not formalised (as in Tyne and Wear's case, see below), but the regular monthly meeting of public and quasi-public sector officers in the conurbation formed a negotiating arena for the interests of the agencies, and decided the broad parameters of the strategy. The only formal process which did exist in this context was the regional submission to the European Commission for structural funds and, again, all of the agencies participated in an officer forum to develop the European strategy.

The Trafford case study illustrates a professional alliance operating a strategic decision-making framework in the sub-region and local specific clusters with a greater mix of public private interests feeding into this. There is considerable continuity of local actors, but this operates at a very local level. Beyond this a series of new networks is forming, bringing in new members to the decision-making arena. But again back-bench power is diminished and the key actors have become public and private sector professionals.

Representation and Accountability

Trafford Park continued competitive working arrangements with fellow agencies and local authorities, but developed a consensual officer framework and sought consensus with community and political interests. Trafford Park UDC tended to argue with central government over financial matters. Coordination was quite important for Trafford Park, this was especially due again to the presence of an overarching European framework. Coordination is less important to Trafford Park because as an agency it only really operates in one area, it does however have cross are projects.

Trafford Park local authority always had good brokerage and negotiation skills and like Tyne and Wear was fairly entrepreneurial. Trafford Park is promoting a new economic development function and internal competition for excellence in training and product development. It is also supporting the shift in the concentration of state resources from the inner city to the assistance of international capital and a move in the decision-making structure towards a non-democratic arena, where representatives of local authorities are excluded. It is possible to suggest that Trafford Park Development Corporation is part of a new alliance of urban officials and senior public figures who are creating the foundation for a new coherence. This process is almost completely void of democratic representation and its legitimation is based upon the results it produces in terms of development or indeed the ability to competitively bid for international events, as with the Olympic Games. The mixture of competition and consensus arising from the very clear spatial differentiation in the functions of the conurbation only gives rise to conflict when there is a difficulty about attempting to redefine the territorial division which has more strength in the informal arenas that it does in the Unitary Development Plan. The UDC has firm control over the access procedure. The inetactio between agencies is divided into two levels. One set of interactive mechanisms operates at sub regional level. This involves a network of professional in governmental and quasi governmental bodies and although councillors do have some input into these joint fora, this is limited to senior elected figures. Another set of mechanism operates within the park itself. This is to do with local forms and the UDA and local authorities. Where there are formal issue based meetings between the UDCs and local companies or regular meetings with local authorities.

New Policy Processes

In most cases the processual framework adopted in the post-UDC era shows only marginal difference to that adopted prior to its establishment. This is largely due to the continuity in officer personnel. If anything there has been with one exception a more general trend in order to create a more consensual environment. As I state above this appears to be due to the overarching framework of the European supranational state, which induces a cohesive regional planning environment. However when we move onto the policy style, there is some differentiation which can be attributed to structural institutional imperatives as well as a transformation in the organisational capacity of the state. Here there appeared to be considerable continuity in the values, ethics and skills of UDC personnel. But the rules of engagement, the approach to opposition and consultation and the control of implementation mechanisms have undergone significant change. Interestingly enough this sits alongside a preferred decision-making mode that could be ascribed to the Keynesian state era.

The framework developed in line with government wishes. But again when we come to the more discretionary elements i.e. the structure of competition and collaboration we see the influence of a more coherent group of officers who have their own agenda.

Figure 5.3 Decision-making Frameworks

Framework	**Trafford Park Local Authority**	**Trafford Park Urban Development Corporation**
Desired planning framework	Strategic	Strategic, flexible/ responsive
Desired internal organisation	Departmental (small)	Small teams
Desired personnel practice	In-house and use of consultancies (private sector)	Local authority offices and private sector consultancies
Structure of competition	Consensus with local private sector interests. Competition with other LA districts rather than with central government	Competition with one district: Manchester, though co-operation with other districts and with local interest in private sector (i.e. Trafford Park competitive with central government)
Structure of co-ordination and collaboration	Inter-departmental Intersectoral Interagency	Inter-organisational agency. Intersectoral Regional
Desired interagency interface	Hierarchical. Attempts to be strategic	Strategic, hierarchical but fluid and ad hoc as well

The decision-making style is more difficult to categorise as the boundaries between a Keynesian and new Keynesian mode appear much more fluid. It seems that rather than 'using' their powers of control over the decision-making environment, Trafford Park appear to keep them in reserve.

Figure 5.4 Decision-making Styles

Style	Trafford Park Local Authority	Trafford Park Urban Development Corporation
Dominant value	Public/economic good	Economic and public good
Dominant ethic	Professionalism and entrepreneurialism	Entrepreneurial and public sector professionalism
Dominant skills	Managerial opportunist	Managerial and brokerage
Attitude to competition	Within local authority and between districts	Inter-organisational with occasional inter-regional
Approach to development	Speculative and risk taking within a loosely co-ordinated framework of development plan	Co-ordinated approach – occasional risk speculation
Rules of engagement	Rules clear, engagement, contingent all part of process	Rules clearly spelt out. Minimal access all contingent on permission of UDC
Legitimation of economic development planning	Legitimation of economic planning decisions on 3 'E's and measurable primary financial criteria	Legitimation via 3 'E's and project success
Approach to opposition	Co-option of opposition. Stabilisation through strongly limited interests	Delegitimation over exclusion of opposition though limited co-option used
Approach to consultation	Informative	Post hoc informative
Control of implementation mechanisms	Control of provision	Regulation of external groups
Desired decision-making mode	Rational	Pragmatic rational

6 Tyne and Wear Development Corporation

Introduction

Tyne and Wear Development Corporation was created in 1987. It covers 2,375 hectares, over 27 miles of riverside, the Tyne in Newcastle and the Wear in Sunderland. The UDA encompasses four district boundaries, Newcastle, Sunderland, South and North Tyneside. Gateshead is the remaining Tyne and Wear district (1).

This case study is limited to the north bank of the Tyne, the City of Newcastle and the Metropolitan Borough of North Tyneside. Therefore, it concentrates on the relationship with selected agencies and actors in the locality of Tyneside.

Announcement and Designation of Tyne and Wear Development Corporation

The UDA's population is small at 4,500. The boundary was drawn to deliberately minimise a resident population , thus removing any unnecessary interference in Corporation policy. Like both Trafford and Bristol, the UDA contained substantial employment 40,115 (Price Waterhouse 1987). At the time of designation the Corporation's budget was £160 million over ten years but in common with all of the other Development Corporations their budget has since expanded ((DOE 1993).

The UDA is larger than the original direction area given by the DOE. Three changes were made to the initial plan. Firstly, the inclusion of a wider area in

Newcastle Quayside, pushing toward the city centre this gave greater depth to the UDA and allowed more scope for the UDC to create viable development sites and exploit the City's regional capital and commercial status. Secondly, the inclusion of two run down industrial estates in Sunderland opened up further potential development sites, and thirdly the exclusion in South Tyneside of areas of housing around Bill Quay, Hebburn and Jarrow which were already subject o development and diverted the UDC from its preferred strategy in this area (Price Waterhouse 1987). The area benefits from two Enterprise Zones. One was already in existence when the other was designated and includes the Newcastle business park in the west of the City. The zone had a ten year life and was designated in 1982. Sunderland gained a new enterprise zone, now named Sunderland Enterprise Park, on the demise of Sunderland Shipyards in 1989.

The Local Political Framework

The Tyne and Wear conurbation is not as politically fragmented as Manchester or Bristol. It is made up of five district councils all of which are controlled by the Labour Party. Little remains of the previously active Conservative Party on Tyneside. In addition to the District authorities Tyne and Wear County Council existed until in 1986. Tyne and Wear was also Labour controlled, though it was never popular with the lower tier councils. The County Council often mediated the local rivalry between the authorities, especially in the field of economic development where it performed a strategic policy and research function in an attempt to create more coherence within the area.

The labour movement, particularly the General and Municipal Boilermakers Union and the Transport and General Workers Union maintains a powerful position within the area, often through sponsored members of local authorities. The umbrella body nominally controlling these two elements, co-opting a third, the Northern Group of Labour MPs, is the Regional Labour Party. The Party, as early as 1978, began to worry about the response to economic restructuring. At the time they sought a resolution to the problem through regional corporatism, blaming the problems on 'regional tribalism', the competition for jobs and investment that existed between authorities (TUSIU 1988). In 1981 they published "A Future Together" which suggested that all organisations in the region should pull together to

> 'provide the framework within local government, industry and commerce, the TUC and English Estates and The North of England Development Council, and the regional offices of central government can more effectively develop and implement agreed policies and programme for resolving the full range of industrial problems within the region' (TUSIU 1988).

The level of political integration in Tyneside (as compared to Bristol, see above) was immense, and although it was factionalised and 'tribal', its sub-regional cohesiveness was stronger than in Manchester. This was temporarily threatened in 1978 by the publication of the Alternative Economic Strategy by the Conference for Socialist Economists and which percolated through to the Regional Labour Party (CSE 1978). Amongst other things this proposed an alternative approach to economic development for local authorities, but was more enthusiastically welcomed by those outside the Labour Party in the region than inside (2).

Politics within municipal authorities echoed those of the regional Party. Newcastle City Council shifted from an authoritarian right wing labour group in the 1960s to a centre right coalition in the 1980s. The administrations of the mid seventies instituted a number of innovative measures in council services, especially in economic development. During the 1980s group tensions spilled over into splits over issues such as budget reductions and eventually rate capping. At this time a small 'urban left' group expanded into a much larger coalition of centre left activists most of whom had been brought into the Labour Party on the back of the new social movements, particularly the Peace and the Women's movement. The leader was the same for the entire period and maintained his hold upon the ruling group throughout the 1980s, carefully crafting the introduction of new members into leading positions but maintaining the balance of power through the old guard.

Beyond the Council, however, active tenants groups, community ownership movements, and small groups campaigning for better facilities within local estates became more visible. The groups were predominantly female, had links within the council, but remained independent (Campbell 1993). The groups were neither homogeneous nor necessarily ideologically coherent but they became entangled in the council's decision-making arena, whether they were asking Economic Development Committee to pump prime a credit union, a priority area team to fund creches or the Policy and Resource committee to back a call to develop a European funded training centre for women. They were crucial in pushing the council and then the Urban Development Corporation to extend its concept of economic development activity.

Urban Policy and Economic Development in Tyneside

Tyne and Wear authorities have enjoyed Urban Programme status, either as Partnership authorities or as programme areas, since 1978. The majority of the programme finance is spent on the Partnership districts of Newcastle and Gateshead. As the eighties progressed expenditure on the urban programme diminished while expenditure on UDCs rose (Robinson 1989). Between 1979 and 1989 resources were also redistributed within the programme. Economic development activities took an increasing amount of the expenditure (Robinson 1989).

The area however, is no stranger to urban and regional policy initiatives, they have played a key part in its governance since the 1930s. The coincidence of economic decline and high levels of poverty has meant that Tyne and Wear has long participated in government programmes designed to deal with the social and economic effects of restructuring. This has created a peculiarly pragmatic culture of central local relations whereby the authorities have been intensely proactive in developing initiatives designed to perform the task of regeneration. This relationship reached its apogee at the beginning of the 1980s.

Latterly local authorities have been extremely active in economic development and inner city regeneration policy. In addition to the local authorities, a profusion of agencies committed to urban regeneration were established in Tyne and Wear during the 1980s. This was particularly apparent after the demise of the County Council.

The Northern Development Company emerged in 1985 to become the regional (corporatist) inward investment agency (3). The Northern Councillors Association (the agency's political equivalent) was a less ambitious body made up of representatives of all the region's local authorities. It formed in the wake of NDC and was seen as the publicly accountable element of the NDC framework. In 1988 The Newcastle Initiative was launched and The Wearside Opportunity was established in Sunderland. These two organisations represented the move towards public private sector partnerships which emerged from the CBI publication, 'Initiatives beyond Charity', and the activities of Business in the Community (CBI 1988). The new agencies brought together many of the members already active in a range of agencies including NDC and the Development Corporation. The Tyne and Wear Development Company was established to manage the property development element of the late County Council's work and by 1987 the City Action Team was also operating in Newcastle City. By 1991 there were also two City Challenge areas, in North Tyneside and in Newcastle.

Within the conurbation Newcastle has always stood to gain most from its relationship with central government. This meant that by the early 1990's, despite the chasm between central and local government generally, Newcastle was adept at securing funds or strategies designed to benefit the City. It established an Economic Development Unit in 1976. The unit was housed in the Chief Executive's department and concentrated on developing small businesses, incubator workshops and giving business advice (City of Newcastle 1983).

The City was influenced by local and national campaigns in the early 1980s, and developed an interest in pursuing alternative employment and ownership strategies (Wainwright 1989) which were being promoted at that time by the so called new 'urban left' authorities. By the mid eighties the culture in the department had reverted to corporatism. After the 1987 election the unit grew significantly and, though its budget was small in comparison to the major spending departments (£2 .5 million out of a net budget of £200 million in 1987 (Newcastle City Council 1988), its strategic importance had grown within the authority and its functional remit included the most significant corporate responsibilities. By 1990 it had swallowed the much bigger Planning and Estates Departments to become the Development Department. This department pursued Newcastle's role as regional capital (Newcastle City Council 1991).

Preceding Strategic Framework

The UDC was therefore plunged into a traditionally active planning arena. The Northern Region Strategy Team published the Strategic Plan for the Northern Region in 1977. This was written as part of the ongoing campaign originated by Councillor T. Dan Smith (who established the Regional council for the North in the 1960s) for co-ordination of the regional economy and was incorporated into the later Tyne and Wear Structure Plan. The Plan suggested that the region had not benefited from the multiplier effect on local service industries that was thought to accrue from inward investment. It seriously questioned whether the growth in office and professional services in the region would allow the region's capital to become a national centre for business services which some had advocated as a solution to the region's economic catastrophe. It advocated the provision of light industrial and commercial premises across the conurbation, with an increased emphasis upon developing sites in older industrial areas and more flexible planning policies for the introduction of more light engineering and manufacturing industries. Finally it advocated environmental upgrading, stressing the rehabilitation of older industrial sites (TUSIU 1988). In may senses it predicted the strategic direction of the UDC. And like the UDC it articulated some of the dilemmas of economic development activity in an area until very recently almost completely colonised by heavy manufacturing industry.

The emphasis of the Tyne and Wear County Structure plan, like that of the preceding plan was to direct investment towards inner city sites and industrial areas. It promoted major derelict land schemes, tightened the operation of the green belt, and emphasised retail development in existing centres. Office development was directed towards Newcastle City Centre, and industry was directed to the south of the conurbation.

Three local planning documents cover the Newcastle UDA in addition to the Tyne and Wear County Structure plan – the City Centre local plan, the Regeneration of the Quayside Project and the River Tyne Subject Plan. These are remarkably similar to the strategies which have been adopted by the UDC. The document which has now superseded all of these is the Unitary Development Plan, which went out to public consultation in May 1991.

The Newcastle Unitary Development Plan was contentious and fuelled acrimonious debates within the ruling Labour group of the City Council. It proposed significant alterations to the Structure Plan policy, redirecting new development to peripheral areas and invasion of the green belt for residential uses. It promotes inward investment and image building strategies linked to the development of central infrastructure. Within the UDA, the proposals of the UDP concur entirely with the Urban Development Corporation's views, emphasising the development of commercial, leisure and recreation uses on the Quayside and a major commercial, leisure and recreation site west of Central Station (TWDC 1991).

Only two of the North Tyneside local plan areas were affected by the designation of the UDC, the Wallsend and Willington Quay local plans. However, Royal Quays is not covered by either of North Tyneside's local plans and in the arguments which took place over the uses for the site, local communities and the Borough Council drew upon the Tyne and Wear Structure Plan – specifically in relation to the guidance on retail sites – to argue for a reduction in retail space.

The Urban Development Corporation

Establishing the Board and Management Framework

The Board of the Development Corporation has twelve members, eleven men and one woman. There are three councillors and one former Leader of the Conservative group on Newcastle City Council. The private sector members represent industrial capital, by and large, though three of the board members have commercial and finance capital backgrounds. There is one academic on the board. The firms represented combine a mixture of local and international organisations. What is interesting about the board is the linkages between its members and other quangos, public/private sector partnerships and public bodies in the region, and in many cases their mutual memberships of many boards. It is a board that is deeply embedded in the socio-political relationships of the region.

Others have noted the very strong horizontal networks which operate within the region (Shaw 1990). At first glance the area appears to have maintained a corporatist identity, a strong trade union influence, a strong labour movement, significant public sector influence and a handful of large local companies span the breadth and length of the decision-making process. In acknowledging the presence of all these characteristics in the Board of Tyne and Wear Development Corporation, it is important to make clear that the institutional framework from which they spring is a framework that is increasingly formally fragmented and unaccountable.

The officers of the Corporation have been said to operate using a public sector ethos (Robinson et al 1993). Their peculiar ability to get on with the local authorities in the area has been attributed to the Chief Executive's experience as the previous Regional Controller of the Department of Environment and Transport. Alistair Balls central government experience, particularly in developing the formula for devolution under a Labour government in 1979, gave him a generous lesson in the intricacies of regional politics. But perhaps what has been more useful in this context is the work which he participated in around the establishment of the Northern Development Company, to which he was clerk. In fact Balls' 'watched the network being created' and formed an invaluable bank of contacts, because he was company administrator. It is true that many of the senior members of staff not only have a public sector background, but almost the entire executive team are Scottish. Robinson says of them:

> 'There is a business veneer based on deals in land and property, while underneath the philosophy is perhaps not unlike that of a local authority' (Robinson et al. 1993).

This perhaps underestimates the complexity of developments in the arena of urban regeneration. When headhunting staff the Chief Executive, apparently searched for people who could 'strike deals'; who had 'hunger and grit'. What he did not want were 'people from New Towns' because 'while they've got the development orientation, they have no sense of the political environment'). He did not therefore look for staff in local authorities, nor in New Towns Development Corporations, but at the Scottish and Welsh Development Agencies and the other UDCs; in other words the new agencies of regeneration which had grown up since the seventies. He found staff who could think beyond departmentalism and closed professional boundaries, people who could strike deals:

> 'There is a skill which has come together, that understands (regeneration) far better, which I don't think existed before'.

The Corporation is not clearly classifiable as it is not quite public and not quite private sector as having an entirely new range of skills that suits the new agencies.

After the formation study, further strategic work was done by the Corporation. The UDC sought to 'take everybody' with them, in the words of the Chief Executive, this was done through the 'Vision 2000' exercise). The document did two things: it began a dialogue with the local authorities in the area, but, in addition, it flagged up where the future difficulties might lie. In some sense the Vision was not only a statement of aims and objectives, it was also a way of testing the political waters, signing authorities up or marking them out:

> 'It was designed as a way of getting everybody...to agree what was an appropriate plan. We took the local authorities with us up to a point. It certainly meant that we had no difficulties in Newcastle. They were broadly happy with us. North Tyneside were always going to fight us on the industrial issue. In Sunderland we found ourselves getting the backing of members and not officers' (*Interview* 12).

South Tyneside was not mentioned and interestingly the initial responses of the authorities seem to correlate with the amount of development that has taken place in the districts. The Vision document is interesting for two reasons. It tells us something about the philosophy of the Corporation and it allows us to compare initial thoughts with later strategies. The document emphasises the role of local authorities less than any subsequent document which the Corporation has produced. The Partnerships, of which the Corporation speaks, are those it wishes to develop with the private sector, with developers and with the Northern Development Company. It is not so much a land use plan as both an extensive and intensive model of the economic capacity of the conurbation. It not only details the site developments and flagships, but it proposes its own and, in some cases, the existing local authorities' preferences for the future economic profile of each district. In this sense it imposed a district wide strategy in each of the boroughs:

> 'The Corporation's vision for the future allots each district a distinct and yet interlinked role in the overall strategy of business/industrial diversification' (TWDC 1989).

The document tells us what these roles should be:

> 'Newcastle is naturally the region's capital, with a strong growing professional and commercial core. Sunderland has the skills and pragmatic tenacity to become the advanced manufacturing centre of the North. North Tyneside can develop traditional fishing and manufacturing industry while developing its natural tourist attractions. South Tyneside already has clusters

> of manufacturing..it has, like North Tyneside, opportunities for sport recreation and leisure and finally both North and South Tyneside support their larger neighbours with labour and services' (TWDC 1989).

Two points occur in relation to the local authorities' response to these plans. Firstly as long as they were consistent with the authorities' own vision, there was less likelihood of disagreement. Secondly, it is surprising that the local authorities did not object to this sort of intervention by the Corporation.

The Vision statement contains detailed sector studies as well as what roughly approximates to a strategic analysis (SWOT) of each district. The impression is of a comprehensive plan detailing not only the general direction of the district's activities but also the strengths and weaknesses of relative sectors within each area. The target sectors for Corporation activity are clearly laid out in the document, it is a detailed economic plan for Tyne and Wear. Far from the Development Corporation being a non-interventionist anti-planning body, it is highly interventionist and the maintenance of almost all the strategies to date must attest to one of two things, either the Corporation got it right or there is a degree of inflexibility on their part.

Byrne has spoken of the TWDC as 'the prime deindustrialiser of the Tyne' (Byrne in Imre and Thomas 1983). But this appears to be an oversimplified description of either the Corporation's intentions or its actions. It sought to undertake a transformation but incorporated an industrial element using sectoral studies as a means of targeting specific production industries. The production activities would be underpinned by consumption strategies dependent on the twin axis of shoring up training in service industries and providing the quality life style associated with new consumption strategies, the development leisure industries, environmental improvement and cultural development.

Additionally, however, the Corporation proposed an image reformation of the area, again dependant on a twin axis, an attempt to boost the external image and the confidence of the region by operating an inward investment drive, attracting new industry and new uses, but also promoting change within – 'the opportunities for local entrepreneurial action, given the dominance of engineering, shipbuilding and coal, were..limited'. Their intention was therefore to 'widen the industrial base'. The Vision document therefore plays both a bold and a delicate hand, laying out the UDCs vision for local districts but juggling at the same time the conurbations' adherence to its history with reflection on the means of economic departure:

> 'We will commit considerable efforts and resources to ensure that divisions between the old and new do not become established in physical and social terms' (TWDC 1989).

Although the strategies in some cases remained fragmented, decision-makers have interacted through the variety of bodies at a senior political and administrative level in the conurbation. The character of the relationship with each of the districts has been different, though there is a common formal organisational framework. All of the authorities have agency agreements with the Corporation. The councillors meet on a formal basis with the Corporation once per year, while there is a monthly chief executives, meeting and regular day to day contact between officers at a lower level. The Corporation sought to form alliances with local authorities.

In Newcastle the relationship has been good. Most of the conflict in Newcastle has centred on other agencies and local MPs. Officers in both the City Council and the Development Corporation perceive their task of regeneration to be a common cause. Newcastle have always maintained that they would have undertaken similar strategies to the Development Corporation had they received access to the same finance and the Corporation have developed or participated in many of the City Council's original schemes (*Interview* 12). Apart from a disagreement over the Science Park, the only other significant criticism of the Development Corporation was when the Leader of Newcastle criticised the Corporation for piecemeal development that lacked an industrial perspective (*Evening Chronicle* 2/8/92).

The most severe criticism of the UDC has emanated from North Tyneside MBC which rejected the strategy for the Royal Quays but which nevertheless retained a working relationship with the Corporation during the entire period of public disagreement; officer links and personal contacts between members are amicable. Since the Deputy Leader, a public critic of the UDC, became the Member of Parliament for Wallsend the difficulties have eased somewhat. This authority's links with organisations affiliated to urban left critics of UDC policy have had a negative impact on relationships between the UDC and North Tyneside.

Both Sunderland and South Tyneside are less bound up with the Corporation. South Tyneside are most critical of the UDC, because they feel they have received the least benefit from the Corporation's programmes. This is reinforced by the attitude of officers in the Corporation to South Tyneside:

> 'South Tyneside is very like Sunderland but more remote' (*Interview* 12).

At a strategic level, when dealing with the leadership of an authority, there are quite clear parameters. At the local and ward level it is not so clear. Backbenchers much more so than their front bench colleagues were found to be in search of a role under the new arrangements, especially when it came to community development.

The cultural differences between local authorities in the North of England and those in the South became immediately apparent and illustrates that the cultural milieu of the Development Corporation was neither monolithic nor entirely reactionary.

> 'London authorities were for the most part very radical – they talked about women's issues, they talked about gay and lesbian issues and coming up here, I just couldn't believe it at first they all seemed homogeneous, old fashioned Labour party, white middle-aged male pragmatists, as time went on it became apparent that they were different' (*Interview* 14).

Despite their differences, however – North Tyneside for instance being 'much younger, open minded' or 'Sunderland- the stereotype' and 'Newcastle a mix of both with the consequent friction' – all of the authorities grew increasingly suspicious of the work that the Development Corporation was undertaking in Community Development. This was partly a case of territorial rivalry but there were other issues at stake as well:

> 'It is also about specific use of community development when you are winning hearts and minds in a political sense and I think they feel very strongly that if anybody is going to do it with their communities it is going to be them' (*Interview* 14).

Another issue which was raised in the context of this relationship is the local authorities' relationships to their own communities. This seems to be a matter not only of the communities being won over to the Development Corporation but about the local authorities' suspicion of their own community groups. Even in the Borough of North Tyneside, the inability of the authority to let go control of the community is apparent. One Councillor suspects the leading people in the monitoring panel to be 'working class Tories'. Asked if there was a suspicion by the authority that TWDC had carefully selected the members of the monitoring panel, he stated that he had no evidence of this but hinted that TWDC was not above exploiting

> 'different groups rivalling each other for a share of the money' (*Interview* 15).

An officer from the authority always attends the monitoring panel and occasionally focuses the panel on 'tactics' but

> 'the monitoring panel is not at all led by the council'. In fact, 'the monitoring panel is viewed with disdain by the council as a whole' (*Interview* 15).

If that position pertains in North Tyneside, the logic of the politics of the area would dictate that it is even worse in Newcastle, so much so that developments in the east of the city, which the councillors did not approve of, were subject to a coup by local councillors. The Development Corporation's position is delicate though clear:

> 'we are not going to do it if it is going to go through this councillor led group' (*Interview* 12).

The issues for local authorities have not therefore developed along simply party political or ideological lines, except to a certain extent in North Tyneside. They have been about the control of territory and the ideas about what type of development is considered appropriate for certain sites. Issues about accountability have not had the strategic importance that they held in Bristol. On the other hand, when it comes to the local members working with community groups, accountability has assumed far more importance, perhaps for the reason that these members do not have the access to the same informal networks as their leading members.

Most of the public opposition to the UDC has been led by or involved local Members of Parliament in the area. The member for Newcastle East has and continues to be a strident critic of the Corporation. The current member for Wallsend, has attacked the Corporation most recently for the level of grant which they receive, contrasting it with reducing expenditure of local authorities. The previous member for Sunderland North remains a critic of what he sees as the Corporation's inaction in the face of the threats to Sunderland's shipyard industry.

In fact the Northern group of MPs have turned out to be far more antagonistic to the UDC than local councillors, whose powers have diminished as a result of the creation of the UDC. Perhaps the reason for this, aside from their role being a far more specifically political one in the context of central local relations, is that their role has become far more limited in relation to the decisions that are being made about their constituencies. The councils, as we have seen, still have considerable potential input into the decision-making process of the UDC. Whereas in an area like the North East, Members of Parliament were previously quite involved in council decision-making through the networks that have been explained above, now have less control over development and its impact. One of the key political impacts of the UDC has been to further neuter Labour MPs and to force them to adopt the tactics of groups that in the early eighties were external to the established decision-making structure of the region, namely waging press and public information campaigns against the incumbents in power.

As we have seen elsewhere, a critical feature of the capacity to develop an opposition to the UDC has been the use of both national evaluation studies of their impacts, and national networks which have been involved in monitoring the Corporations. Both Councils are affiliated to national networks. It is significant that the Corporation has used national evaluation studies in developing their framework for operation, particularly in developing a community and training strategy. The Corporation has suggested in interviews that the 1988 Report of the Employment Select Committee allowed them to become more involved in the non-physical aspects of regeneration. Furthermore they suggest that it helped them in defining their own approach to regeneration because it gave more guidance on what the word meant, the 1980 definition contained in the Act being too wide.

Community Development

Tyne and Wear's community experiments, in the words of a previous critic of the Corporation, 'constitutes best practice' (Robinson et al. 1993). The philosophy of developing close links with communities became evident at the time of the publication of the Vision strategy. Tyne and Wear, like other second and third generation Corporations, had seen Docklands and had not been impressed. So although Tyne and Wear was initially wont to talk about the west of Central station site as being a 'Canary Wharf', within a few pages they were proclaiming:

> 'We do not want to create glossy multi million pound developments on the one hand and ignore areas of social need on the other. We will take steps to avoid division... and aim to create harmony rather than conflict' (TWDC 1989).

The Corporation, from a relatively early stage, details a comprehensive community development strategy, which provides the foundation for consultation mechanisms, for involving people in the design of new facilities, for supporting self-managed housing schemes and developing local employment contracts with developers. They made clear their intention to support small business and local economic development initiatives and they contracted out their grant giving to a local charitable organisation.

Since this original commitment, community development has grown: firstly in terms of its budget, from £252,000 in 1988 to £1.7 million in 1991/2; secondly in the size of department within the Corporation; and thirdly in the scope of its activities. The original commitment came from a policy document entitled 'Strategy for Community Development' (TWDC 1989b). This anticipated potential problems for the Corporation as a result of the juxtaposition of 'the most deprived communities in the North East' and the Corporation's boundaries. So, although there was only a small population within the UDA, the UDC was far from complacent about its position. Much of the sense of the first document was drawn from the initial community consultant's meetings with community groups. The strategy produced guidance on the approach to consultation, the type of support that would be given to communities interested in community or self employment, the investigation of training schemes, exploring a range of social housing options, involving residents in the design of their own environment and the criteria for funding community groups.

The community development strategy has been revised twice since 1989 and has become much more strategic. Community Development and Training were amalgamated in December 1989 as it was felt by the Corporation that training was a priority for community groups. Both the training and the social housing programmes have advanced far beyond the agreements developed elsewhere. Social housing comprises 25% of the new housing stock built by the Corporation (TWDC 1994-5). A joint training strategy with the Tyne and the Wear Training and Enterprise Councils, funded up to £5 million pounds by the Corporation, has been produced. The strategy is being implemented with the assistance of three secondees from the partner TECs and has developed a wide range of training activities (TWDC 1991-92). The bulk of the community development and training now:

> 'sits firmly alongside property development. Now it's about paving the way for development and enabling people to influence development' (*Interview* 14).

Within the context of community development, perhaps most attention has been paid to the Monitoring Panels which have been established in East Quayside, Royal Quays and North Dock and Downs, the three most difficult sites for the Corporation. The monitoring panels have very different characteristics and have evolved in different ways. East Quayside was the first to develop. It arose from discussions raised by local church groups, in the east end of the City – the East Quayside group. Residents, workers and councillors in the area were worried by the number of small local organisations that were likely to be affected by the Development Corporation's proposals, such as the pigeon fanciers and the Ouseburn water-sports club. A local councillor was interviewed by the *Sunday Times* where he poured contempt on the Corporation's lack of foresight. The monitoring panel was conceived at a meeting where the Corporation 'tore a strip off' him (*Interview* 16).

The Corporation suggested that a panel be set up and that he nominate some local people, but then everybody got involved. Like the other monitoring panels, meetings take place on a monthly basis, with a planning meeting for the officers of the group in between. Developers and three Corporation staff attend. Each panel has an administrative budget of about £6,000. Inevitably, apart from a regular report back on the progress of the development, training is top of the agenda. Social housing, private housing and infrastructure are also discussed. East Quayside is seen as a complete anomaly by the Corporation as it is chaired by a local member.

The other two monitoring panels are seen as far more community based, with local members having much less control. When asking the Corporation which of the groups had influenced them most, they felt it was Royal Quays, and that East Quayside probably influenced them least of all. They felt that this was because:

> 'It is easier to give into members of the community than it is to give into politicians' (*Interview* 14).

Of course the types of development has also made a difference: 'North Dock/North Downs has always had more flexibility about it – East Quayside has always been much more difficult in every sense and much more high profile which makes it difficult to accept a high degree of influence. In Royal Quays they did have a major influence on the amount of social housing, the nature of the waterpark and the amount of retail and commercial space (*Interview* 13).

The Urban Development Area

There are four flagship sites within the UDC, two on the north banks of the Tyne, East Quayside and Royal Quays, and two in the Wearside area, Sunderland Enterprise Park and St. Peters Riverside. The latter two will receive a more cursory treatment within this chapter. In addition to looking at the flagships, however, I will range over the prominent sites throughout the UDA, as a number of smaller disputes have been significant in shaping the dimensions of local social relations.

By 1992 the emphasis upon Flagship development was the most prominent force in directing the expenditure of the UDC, 53% of the expenditure in 1991/92 was on Flagship development (Robinson et al. 1993). Although it is argued by Robinson et al that this will mean that the concentration of the Development Corporation's activity will remain on the north banks of the Tyne, particularly in Newcastle, the Annual Report for 1991/92 suggests a lot of movement around the Sunderland projects. The concentration on Flagship development in the UDA, to the extent now that Community Development and Training programmes are synchronised and funded in accordance with them, has been developed because of the short time scale of the agency's life and the need to maximise development on the ground as part of the output funded approach by central government.

During the 19th century the riverside hosted the great industries that propelled the region into unparalleled economic power in both shipbuilding and engineering (Byrne 1993). Throughout the second half of the 20th century the decline has been speedy. Recently, the most significant employment losses in the

Figure 6.1 Tyne and Wear Urban Development Corporation: The Site

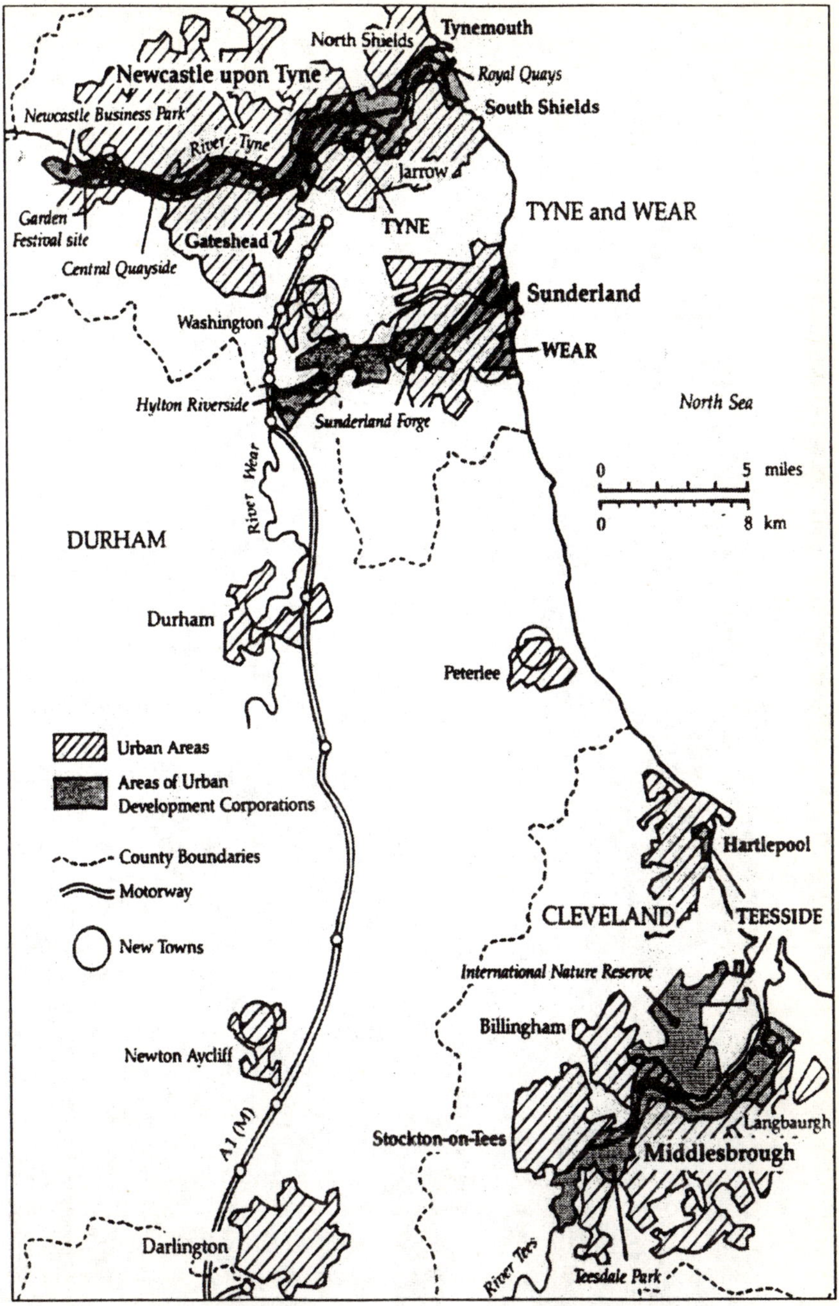

Figure 6.2 Tyne and Wear Development Corporation UDA

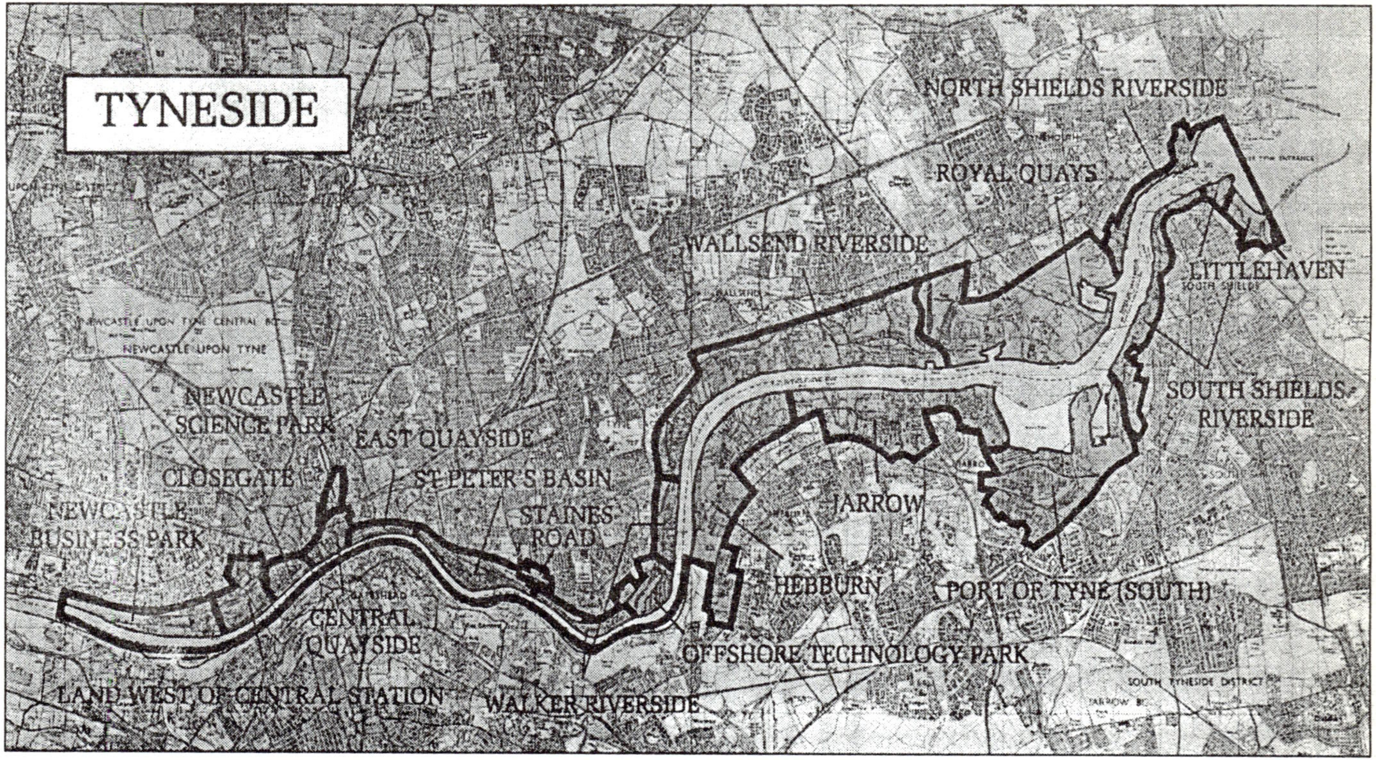

west of Newcastle were the result of the 1980s restructuring of the multinational Vickers Group. This meant the removal of Vickers workforce from Elswick westward to the Armstrong works. This was predicated on the partial closure in Elswick of the foundry in 1980 and the non-ferrous metals sheds in 1981 (TUSIU 1988). The derelict site was later bought by the City Council. The Scotswood factory of Vickers closed in 1977 and a new factory Michel Bearings was rebuilt soon afterwards, the cost borne entirely through Enterprise Zone inducements.

Newcastle Quayside has always been the historic core of the city, it remained untouched until the recent property boom. From the Ouseburn in east Newcastle to the river mouth at Tynemouth, a glorious mix of significant industrial and shipbuilding capacity, off-shore marine technology and small marine engineering supply industries coexist with substantial dereliction in the wake of declining coal export capacity. The now threatened yards of Swan Hunter lie amidst this industrial tapestry.

To the south of the River, the borough of South Tyneside exhibits a patchwork quilt of land ownerships and land uses. These range from new housing development at Hebburn to marine related and off-shore uses up river. But there is a vast amount of dereliction arising from the departure of engineering industries (4). Sunderland, until 1988, was a working port. The demise of NESL meant the effective end of shipbuilding on the Wear. In addition, the Riverside sites showed little promise of diversification because of the steep escarpment at the riverside. Although the borough of Sunderland owns much of the area they have not shown much interest in developing it, concentrating instead upon the greenfield sites which had proven so attractive to Japanese investors.

The challenge to the Corporation was to secure integrated development as a subregional area across disparate sites (Price Waterhouse 1987). The manufacturing base in the area was largely destroyed by the 1981 recession. This left two substantial shipbuilding concerns in the area – only one now exists. The fact that this crisis had all but destroyed an already hugely diminished economic base, meant that many of the strategies introduced by the UDC were already being looked at by local authorities.

West Quayside Newcastle Business Park

This is a 60 acre site in Elswick which was previously occupied by Vickers Armstrong and was purchased by the City Council in 1986. The City Council was at the time in partnership with a local developer, Dysart, to build a retail and leisure led business park. When TWDC bought the site in 1987 it maintained the partnership, but much of the retail and leisure elements were dropped from the scheme and the business park was strengthened. On transferring the site to the UDC, Newcastle City Council stated to the press that it was hampered by the lack of cash to pursue the development and the 'the Development Corporation was in a unique position to inject finance immediately'. The nature of the ensuing relationship between the city council and the UDC was quickly apparent in this transaction.

> 'I am confident that a successful outcome to these negotiations is in the interest of all concerned' (*Evening Chronicle* 24/2/1987)

In fact it was central to the interests of both parties. This relationship was underscored by the words of Nicholas Ridley who on a visit to the UDA made an uncharacteristic comment that the UDC had 'worked in close cooperation with private sector and the local authorities' and 'That's how it should be' (*Planning* 780 5/8/88). The Business Park was 95% complete at the end of 1991 with 90% of the office space there let or sold. The Park is aimed at light industrial or business users. It has since been hailed as one of the UDCs most successful projects (TWDC 1991-92). Most of the infrastructure had been put in place at the time of purchase and has been due to the Enterprise Zone incentives (which ran out in 1991) plus, the Corporation would argue, its own development activities. The park has been successful in attracting new inward investments – British Airways (£36 million/1000 jobs), AA insurance (from elsewhere in the city) £22 million and 250 jobs, Ministry of Agriculture intervention board and others including Merz and Mclellan, IBM, Allied Dunbar, General Accident Property Services, ICL, IKEDA, Technology Europe, and IBM (TWDC 1991-2, 51).

West of Central Station

This site was originally announced as one of the Corporation's Flagship sites which would hold 300,00 square feet of offices, a new concert hall and a technology park (400,000 square feet). There are numerous problems with the site – existing uses, existing landlords, British Rail, own a significant strip of land, an existing bus station which is the basis of travelling into Newcastle from south of the River and difficulties in garnering funding for new uses, for example a proposed CHP station and a major new international concert hall. Crucially, however, the leverage ratios were dependent to some extent on two courses of action being followed by the City Council, the development of the West Central Route which would flow into the site and the decision-making of the existing City Concert-hall, owned by the City Council. More recently the building of the Newcastle Arena with a capacity of 10,000 has been achieved on this site. (TWDC 1994/95).

Forth Banks and Central Quayside

Forth Banks is cut off from the main city but leads to a potentially picturesque area of the old town wall, and some bonded warehouses. A city council reclamation scheme was already in place to develop the wall. On the quayside itself the TWDC attracted the Copthorne chain to build a new 150 bed hotel on a four acre site.

Most of the developments proposed are in line with the city council's preferred strategy for the area. The improvements included a mixed use development, incorporating a pub and a restaurant and flats for rent developed by North Housing Association. Other projects are mixed use and benefit from English Heritage finance.

Newcastle Science Park

This occupies the Manors site next to the Manors metro station, slightly beyond the city centre. It is a 17 acre site and was originally seen as a place for incubator units. This site was the scene of an early spat between the City Council and the Development Corporation. The Council's Economic Development Committee were looking at the development of a multiplex cinema as part of a cultural industries strategy and had identified the north side of the Manors site as the site for the multiplex, because of the competition with the Metro Centre, a shopping mall and leisure complex across the River, in Gateshead. Almost before the scrap had begun, the UDC backed off.

> 'We got into a fight with Jeremy in the early days, it was a mistake, ..it had to be resolved and the solution is ok. We had to come to an understanding with Jeremy, if you are going to develop anything in Newcastle to try to do it in any way in opposition to the city is courting immediate failure and my style has always been to stick as close to Jeremy as I conceivably can' (*Interview* 12).

The disagreement arose from the inadvertent actions of a Corporation employee who just didn't know the rules of operating in Newcastle. After negotiating through the disagreement with the city council over the preferred use for the site, the city council have developed a multi plex 12 screen cinema on the northern part of the site. The Corporation agreed to develop the southern part of the site as a science park with 100,000 square feet of innovation centre, and high tech units between 1,000 and 5,000 square feet. The site is close to the city's Universities.

East Quayside

This is a Flagship site and is 25 acres of land stretching from the milk market to the Ouse burn. The project has been called the 'Jewel in the Crown' of TWDC. The difficulties with this site include a necessary reorganisation of the highway, manoeuvering the difficult topography and dealing with the River Wall. The proposal for the site is mixed use development including:

250,000 square feet of prestige offices
120,000 square feet of catering facilities/speciality shopping
A five star 200 bedroom hotel
A 50,000 square feet exhibition centre
A themed heritage and water-sports centre
325 homes.

This is the most controversial Newcastle site and underwent a three year marathon through the courts as a result of its CPO. There were initially 48 interests in this site and two in particular challenged the Corporation all the way to the High Court over their CPO. Like Wharfside in Manchester the site illustrates the nature of relationships around development in Tyne and Wear.

The East Quayside battle involved a wide number of participants, apart from the UDC itself, which wanted to compulsorily purchase the site for their 'jewel in the crown' flagship. The City Council group members were split about the development. Firstly between the leadership and the back-benches and secondly between different territorial groups.

Many of those who objected to the redevelopment were part of a group of councillors popularly known as 'the east end mafia' denoting their control over activities in the east of the city. There were a variety of reasons for their opposition to the development. All of the members were resistant to non-manufacturing uses and the 'yuppification' of the East Quayside site; they were concerned about their own rights of representation and they were undoubtedly fuelled by the antagonism of their local MP, Nick Brown. Brown, at the beginning of the Corporation's life, had voiced opposition to the UDC, denouncing it as 'the unelected corporation run by Mrs Thatcher's colonial governors' (Journal 11/6/88).

The UDC established a design competition for the site. This approach is not entirely novel in either urban regeneration circles or the political milieu of Newcastle, as a competition to design uses for the Quayside had been staged by the City Council in 1985 (*Interview* 16). The UDC invited three consortia to enter the competition, and each be paid £40,000 to create a model and plans for the East Quayside site. The models were displayed at the City's Guild-hall and through press advertisements the public were invited to attend, view them and state their preferences for the development of the site in a secret ballot. One Councillor was entirely dissatisfied with this, stating that 'the competition itself was an unusual procedure, you usually determine what you want to do first and then choose the developer' (*Interview* 16).

Three developers were involved – Brookmount, Laing and Stanley Millar in partnership with Rosehaugh. The next stage of the consultation was a presentation before the Councils Regional Capital and Urban Development Subcommittee. Before the UDC had selected a preferred developer, the City Council had decided against the Brookmount project. The Chair and Officers at the committee differed with a substantial number of it's members. Some of the East end councillors and certainly the Newcastle East MP had preferred the Laing proposal rather than Stanley Millar. Laing had done a lot of work with the existing housing association, North Housing, and its scheme was more in tune with traditional uses, though even the east end councillors admit that it was 'unimaginative'.

The UDC choose the Millar development, or the Newcastle Quayside Development Consortia. Many suspected they had done so because of the amount of land Millar owned along the Quayside. Millar had shared the ownership of the area when the city docks closed in 1970 with the City Council (*Interview* 17). It was when QDC had been chosen that Procter and Gamble, with Headquarters on the site, suggested that they didn't want to sell their buildings to the Corporation. They had developed an expansion plan for their building and intended to spend £4 million on it. The Stanley Millar development would involve bulldozing the Procter and Gamble site.

A developer, Lesser Landau, then bought the site from Procter and Gamble and gave assurances that he would not redevelop the site. The UDC applied for a CPO of the properties. The site needed planning approval and was therefore put before the Regional Capital and Urban Development Subcommittee of the City Council. Some councillors felt that the Director of Planning's instinct was to turn down the application for the Corporation development, but his report recommended acceptance at the final subcommittee meeting (*Interview*).

At the same time that the Public Enquiry was announced, Procter and Gamble attended a meeting with the City Council. Procter & Gamble were told that there action vis a vis the site was damaging to the City, and that the Council disapproved of their actions. Procter and Gamble's resolve hardened after the meeting, clearly shaping up for a fight.

Later at the enquiry over East Quayside, it is alleged by one councillor that the UDC had made a 'deal' with the City Council requiring it to widen City Road (next to the Procter and Gamble site) to allow the plans come together; this would necessitate the purchase of Procter and Gamble (*Interview* 17). The City Engineer appeared at the public enquiry and supported the UDC over the road, but because he refused to be cross questioned, he simply made a statement that had been agreed by the Councils Highways Committee, any more than this, would be construed as a political pact between the Council and the UDC, triggered the challenge which took the whole issue to the House of Lords.

During this time Stanley Millar collapsed and AMEC regeneration stepped took over the Development. The new Developer, is Newcastle Quayside Developments which is a partnership of Shearwater Properties and AMEC Regeneration Limited. As a spin off of all this controversy, the first monitoring panel was established by the Urban Development Corporation. The monitoring panels are discussed below.

East Newcastle

This comprises the remainder of the Newcastle projects and within this lies St. Peters Basin, which is a new 'village' comprising 289 homes built by Barrett Newcastle Limited for sale and rent. The Callaly Way development in Staines Road, by Cheviot Housing Association, comprises 107 homes for rent, providing family and retirement homes. Walker Riverside Technology Park

This was built by Newcastle City Council and now the Corporation has a partnership with the Council after problems with the site and with the occupiers of the site. The problems include access and landscaping and land reclamation. It was originally the Walker Off-shore Technology Park.

North Tyneside

Wallsend Riverside

This strip of land comprises largely industrial and shipbuilding uses, including Swan Hunter Shipyard, now in receivership, Amec Off-shore and a host of small supply industries. The Development Corporation has supported businesses by giving grants and developing sector strategies. However the Corporation are also keen to promote development opportunities here – 10 acres of redundant railway sidings between Swan Hunter and Davy Lane, which is being viewed as a potential site for inward investors; and 60 acres of land in the Tyne Tunnel area suitable for industrial or warehousing development.

Aside from this the Corporation has grant aided a number of the companies, Charlton Leslie (Offshore) and Press Offshore amongst them. In speaking about this a North Tyneside officer revealed that the conflict between industrial versus non industrial development was not the key conflict, as Byrne suggests. In fact he suggests that the UDC 'expressly and overtly recognise the industrial nature of the area and want to do all that they can to maintain it' (Byrne 1993).

Royal Quays

This is the second flagship and the biggest single regeneration project in the North Tyne area. The Development Corporation have negotiated 200 acres of a 400 acre site owned by the Port of Tyne Authority. The major obstacles to development here were the amount of necessary land clearance and landscaping, wasteland, old foundations, wagonways, discarded engineering works around the 22 acres of docks at Albert Edward Dock. In addition, the opposition by the Port of Tyne Authority, local councillors and local community groups has significantly obstructed the Corporation's plans.

This site has been as controversial as East Quayside. The local authority and community groups were opposed to the original plans of the Corporation, preferring industrial uses for the site instead. The strategy of TWDC was to develop the area as a mini new town for the area, 1,200 homes of which 25% will be social housing units, 320,000 square feet of retail space, 310,000 square feet of industrial space, 10,000 square feet for a training centre, a water park, a hotel, a ten screen cinema, improvements to the 250 birth marina and specialist shopping. Royal Quays has always been a high profile project. In fact TWDC state that it is their most ambitious project. The Port Authority were hostile to the Corporation because of the loss of 200 acres of their 400 acre site.

The UDC had taken a tough stance over the site initially, and was refused its first request to the Secretary of State to just go ahead and vest the land. 'The Secretary of state was not prepared to abandon his own protege' (*Interview* 13). With that avenue abandoned the Development Corporation resorted to 'persuasion pressure, pressure'. North Tyneside was consulted about the mixed use proposal which, in their terms, 'was pretty wild' (*Interview* 18). The community was not involved in the discussions during the early stages. In fact, the joint opposition of the three external parties did not arise until TWDC put its scheme out to tender. Like East Quays, three submissions were drawn up and were featured in a public exhibition. The Port of Tyne, the local authority and the community activists concluded that the balance of the scheme was wrong. It also contravened the Structure Plan, a point that the Port of Tyne Authority was to make great play of, there were not enough industrial uses, and the ability of the Port of Tyne to maintain port related uses within the plan was jeopardised. The amount of retail use that predominated within the scheme threatened the health of an already battered town centre in North Shields. This last objection conflicted with the Corporation's view that Royal Quays should be a New Town.

At this time the local authority began to negotiate heavily with the

Development Corporation. NTMBC wanted manufacturing and industry uses for the site, or alternatively off-shore or improved port facilities. But there was a resonance of political posturing in the NTMBC approach:

> 'It was almost as if we were saying if you don't go ahead with this scheme, this is what we will find acceptable as alternative uses' (*Interview* 18).

North Shields Riverside Action Group was the community element of the response to Royal Quays. Residents of the neighbouring estate in Meadowell and people who worked at the People's Centre for the Unemployed in North Shields developed an alternative plan with North Tyneside, one that was compatible with the Structure Plan and comprised a better balance of industry. It was felt by many on Tyneside that the People's Centre did a reasonable job of debunking the plans of the UDC and the campaign was waged in the local press. Officers in North Tyneside portray a cynical view of the process:

> 'The effect of the debate was actually about positioning, people either being able to claim credit or disown it if it were to go wrong' (*Interview* 18).

Although the members of North Tyneside were of the opinion that the site had always been used for industry, it was previously a coal export site. The north side of Albert Edward dock is where the roll on/roll off facilities are located, but the facilities are badly in need of modernisation. Significantly North Tyneside argue that between the selection of the RQDC proposals and the outline planning application, alterations took place in the plans. The industrial floorspace was doubled, the amount of social housing was increased and fleshed out by a deal with the Housing Corporation and North Tyneside, and a £250,000 training package was added to Royal Quays (*Interview* 18).

What was interesting was that at the time that the very public antagonism within the Riverside Action group arose, publicly the local authority supported the continuing discussion between themselves and the UDC. Eventually North Tyneside asked the Secretary of State to call in the application. He refused, which gave the go ahead for TWDC to begin the project.

North Shield Riverside

TWDC is supporting a modernisation of the local fishing industry and has contributed to the Sea Fisheries Training Centre sponsored by NTMBC. It has given grants to local businesses, refurbished buildings on the fish quay, such as the old Low Lights building, and assisted in the development of a Design Centre on the Quay.

Economic Development in Tyne and Wear

The North East Region, of which Newcastle is the unofficial capital, dips down beyond the River Tees in Cleveland and spans Hadrian's Wall to take in Cumbria on the West Coast. Our analysis has focused upon Tyne and Wear. During the industrial revolution the area was built upon the three industries of shipbuilding, coal and engineering (Hudson 1989). Within the conurbation there has been a clear division of production activities between the local districts (Byrne 1993). Newcastle developed as the commercial and financial heart of the region, but also housed the shipbuilding industries of the Tyne as well as the engineering works of NEI Parsons and Vickers Armstrong. North Tyneside has also traditionally spawned shipbuilding industries and significant port activities, a host of small industries related to the port and more recently off-shore activities. It retains an active fishing fleet and has recently developed a local tourist industry. In South Tyneside shipbuilding and engineering have predominated and only recently has the borough diversified into local tourism, while Sunderland and Gateshead have traditionally been heavily dependant on shipbuilding and coal. Latterly, the Japanese industries of car production and new light manufacturing have arrived in Sunderland with companies such as Nissan and Komatsu. Gateshead has moved into consumption activities, with the development of the Gateshead Metro Centre and the fourth National Garden Festival.

During the post war era, the North East's fortunes have been mixed, but undoubtedly the long term economic trend has been downwards. The rearmament of the Second World War actually disguised the decline that was taking place in the region's manufacturing and engineering companies (Robinson 1990). Regional policy, together with the expansion of the public sector, sustained the economy and gave a veneer of stability until the recession of the mid 1970s. Perilously, however, many of the firms that were attracted through regional aid maintained headquarters elsewhere, and the region was poised to endure the vicissitudes of a branch plant economy.

It was in the the seventies that the North East encountered serious economic trouble. There was a mass exodus of the multi-national companies and, together with the restructuring and decline in many of the region's indigenous industries, this caused the shedding of labour at an exceptional rate. Between 1976 and 1981 43,000 jobs were lost in engineering. In 1981 there were 40,000 redundancies alone in manufacturing. The shake-out devastated the region's employment base. Between 1978 and 1983 the number of employees in the Northern region fell by 185,000 – a drop of 14.9%. Three quarters of these were in manufacturing and they were predominantly full-time jobs held by men (Robinson 1990).

Official unemployment peaked in 1985 at 17.5% but the region has lost

130,000 manufacturing jobs in the last decade and overall manufacturing employment is down by 31% (*Guardian* 14/8/89). While the decimation of manufacturing industry has proceeded apace, there have been some marginal counter trends. The decline has been accompanied by a small growth in the service sector. For instance, between 1979 and 1989 the number of manufacturing employees in the region fell from 410,000 to 281,000, but the service sector grew from 664,000 to 689,000 (Robinson 1990).

The region has a large public sector workforce, the local authorities and the health service employ substantial numbers of people in each of the Districts. Latterly the re-emergence of foreign owned companies has occurred. During the last twenty five years the north's manufacturing workforce employed by outside companies has risen from approximately 50% to 90% (*Guardian* 14/8/89). There is large scale penetration by both Japanese and North American companies. However as Robinson points out, employment in US manufacturing firms also dropped by 40% between 1978 and 1988 (Robinson 1990).

Successive agencies have vigorously pursued an inward investment strategy in an attempt to diversify the region's industrial base and economic structure, promote growth in research and development facilities, and thus secure an upward spiral of indigenous advanced manufacturing capacity. However, despite this and the North's success in attracting new investment, the region's GDP per head remained below all other UK regions except Wales and Northern Ireland in 1989 (4).

The economic trends have consolidated spatial patterns of inner city population dispersal. In Newcastle, for instance, the population declines by 300 every year (Newcastle City Council 1991). Inner city problems have increased with growing rates of crime, consistently high rates of unemployment, infant mortality and morbidity.

There has been a parallel burgeoning of development on the periphery of the main centres and a move in industrial development towards the green areas of Sunderland and the new town areas of Washington, Peterlee and Aycliffe. As part of a county-wide strategy of modernisation and the reintegration of industrial and commercial space, the County Council developed a rapid transit system which links four of the five districts on Tyneside and which was opened in 1981. They also promoted environmental improvement and land reclamation schemes designed to counter the growing swathes of dereliction, the residue of industrial decay (Tyne and Wear County Council 1985). Allied to this, lower tier district authorities pursued consumption strategies. Two major shopping centres have developed in the last twenty years, the first in Newcastle's Eldon Square in the 1970s and the second at the Metro Centre in Gateshead during the 1980s. Tourism too has expanded feeding a proliferation of strategies for environmental improvement on the part of the public sector.

It would appear that the effect of the 1973 crisis was not simply a shift from a fordist to a post fordist regime as substantive remnants of extensive accumulation, for instance shipbuilding, steel and coal industries, coexisted with the regions intensive regimes. It is clear however that in the present period, the region exhibits labour market and consumption characteristics associated with new accumulation methods. These characteristics include technological change, a move to a service economy, flexibility in the wage relation, diversification in consumption strategies and the polarisation of local populations.

It would appear that there is an alliance in the area of which the local authorities remain pivotal players and in which the UDC now plays a full part. This alliance is responsible for propelling both strategic change and change in the form of policy making. It is a long time since the region has exhibited a tendency towards structured coherence. The driving force propelling change towards fordist development, which effectively meant the state underpinning consumption patterns and public infrastructure, rather than any significant changes in production, has been the local state and the organised trades union movement. Indeed in this area of social relations there has been significant continuity as the technological mix has undergone a transformation. In the present period the local state is playing an equally proactive role in grouping together capitalist and previously state monopoly tendencies, but within a new framework.

Thus it would appear that the UDCs are subject to two completely different influences, on the one hand they are propelling the organisational form or culture in a new direction in order to transform the balance of local social relations and the capacity for new technological advances. On the other they retain considerable continuity in the relations around this from, preventing them from cauterising the past.

Representation and the Selection of Controlling Networks

The most significant characteristic of Tyneside, given the roller coaster of economic and industrial development it has endured, is its political and social stability. Despite active trades unionism, there has been a relatively undisputed movement from the industrial emphasis in the region into the commercial and service sector. The main reason for this appears to be the role of the public sector, both through central government regional funds and through the actions of local authorities. The local authorities in the area have sought to do two things: maintain the involvement of the industrial sector in local decision-making, especially at a political level, while at the same time promoting the diversification of the economic base. Because of the close integration of the local authority and the regional establishment, the implementation of UDC policy has been easy.

Although in local authorities the democratic mandate was extremely important in legitimating the actions of the local authority, the provision of jobs in a declining economy was seen as an important task for the economic development units. The success of the project for TWDC was to be measured in terms of its ability to attract investment and 'fill up' its sites and it did not really have to worry about other services. It is notable however that TWDC worked very hard at securing legitimacy from the community and political networks in the area. Tyne and Wear of all of the UDCs was most committed to inclusive strategies even when it came to dealing with the opposition, but even so it was reassured by the option of exclusion should any of its groups become to much of an obstacle to development possibilities.

The people who have power to affect the decision-making structure within the UDC include local authority officers who have been involved in urban regeneration strategies and those who have been involved in the determination of regional policy implementation in the post war era. Front bench politicians, particularly in Newcastle, are crucial to the process operated by the UDC. Experienced community groups have also been able to negotiate access to the UDC in the way that they negotiated access to local authority. Finally there appears to be a network of representatives of industrial, commercial and finance capital who have become involved in the regional partnership initiatives. It appears that there is a tightly knit class alliance, including both public and private sectors. The alliance is based on industrial capital but has been joined by the public sector and elements of commercial capital. The network has extended to include new members, who represent multi-national capital.

Participation and Mechanisms of Accountability

Tyne and Wear Corporation was both consensus-seeking in its relationships with other bodies and competitive over ideas of economic development and urban regeneration. They experienced conflict over the issue of deindustrialisation, but only from certain interests, for instance representatives of the manufacturing trades unions and their counterparts in the local authorities. Because of the nature of the area's politics TWDC appeared to have emphasised good relations with community and political interests. Interestingly the UDCs relationship with central government seems slightly more fractious than that with the local authorities, much of the trouble relating to project development and spending regimes within the DOE. In fact the local authorities in the area had over some years developed considerable links with all of the government departments involved in regeneration and had attracted considerable finance to back urban policy initiatives.

Tyne and Wear managed to sustain a fairly co-ordinated framework seeing itself as part of the region; this was particularly the case when Tyne and Wear County Council existed. Nevertheless there were inter-authority disputes, primarily over the development of prime industrial and commercial sites, everybody for instance not just Sunderland wanted a Nissan, but Newcastle would have preferred it in its own district! Under the UDCs the push for co-ordination was much greater than under the local authority. TWDC was aware of just how important this level of co-ordination could be in terms of submitting European applications for structural development funding. TWDC necessarily had to ensure that the plans for the four local authority areas under its dominion had to correspond.

However the rules of engagement with the UDC in this planning process remain within the domain of a small group of people and it is much harder to get access to the overall decision-making structure of the UDC than it was to gain entry to that of the local authority. Two key sets of struggles have taken place. One was over community benefit, which was as much about the democratic process as it was about strategies designed to benefit communities and the second was about strategic uses in key sites. This was essentially about the level at which manufacturing activity could be sustained in the locality. Conflict occurred after the designation in Tyne and Wear, compromise over the involvement of local communities in the development process around flagship sites was seen by the UDC as quelling local discontent which threatened to break the consensus. Negotiating within the rules of the game seems to describe the process which the leadership and the leading officials from LA's participated in. Negotiating about the rules of the game was reserved for community activists, back bench councillors and the political opposition. Success in negotiating with the UDC was more possible when negotiating within the rules, though not impossible outside them and in this case it can be suggested that the consensus and the strength of the alliance dominated the rationale. Although there has been a gradual process of co-option and incorporation of the various groups in the area, because the local authorities and trades unions still have power in negotiating, there is still a level of bargaining taking place around the UDC. The fact that community channels have now been limited to flagship sites suggests that their power is being curtailed.

Local authorities get the most access to the UDC especially at the level of the leadership. In the case of East Quayside it was crucial that the local authority was on the side of the UDC, but significantly, because of the democratic process within which the local authority operates, it also meant that the process took much longer. Although this caused some problems with land values, because land and property markets are relatively stable in the North, the delay was less significant than it might have been elsewhere. The fact that Newcastle was the commercial capital of the region also assisted the UDC, as the Council was pursuing the same strategy. The fact that most of the key actors had been involved with NDC helped solidify the network.

Opposition in Newcastle came from the representatives of the industrial working class and was based upon differences around lifestyle and leisure uses for the area and jobs and employment strategies. They still wanted traditional manufacturing jobs and expressed a demand for community benefit. The issue of community benefit was raised by an active and experienced church and voluntary sector which was nationally connected to a series of inner city causes. Locally, as a result of long term funded involvement primarily in the Urban programme, the voluntary sector was extremely experienced and educated in negotiating with a paternalistic public sector. The groups who were most antagonistic were those who were outside this traditional sector. In North Tyneside there appears to have been more space for opposition. This was because the local authority was more closely tied into national urban left networks and were linked to labour movement networks. Such networks were inclined to make industrial demands and the councillors assented as they had a very strong industrial base among the members. On the whole these demands aligned with the more prominent new urban left agenda in the authority.

The area always had a tendency towards paternalism. This appears to have become formalised into something like a 'mayoral system' around which a cluster of networks attempt to develop joint strategies within a regional context. Although central state finance is still coming into the area, it has changed form. In Tyne and Wear there is a strong class alliance as a result of public interest domination over the economic agenda which has existed since the inception of regional policy in the 1930's. Although there is a shift in emphasis in Tyne and Wear towards capital rather than sustaining the significant hold of labour representative. There is a good deal of continuity here, but there are also a number of new processes and structures from the monitoring panel to the interagency structure. There is more central state involvement and there is more private sector involvement.

New Policy Making Processes

There has been significant continuity in the policy practices of urban regeneration in the Tyne and Wear area in that there is close correspondence of objectives between the agencies. Much of this is accounted for in the free flow of personnel between the agencies and precipitive actions which Tyne and Wear authorities were taking prior to the establishment of the UDC. Finally it was due in part to the nature of the existing regeneration strategies which were on the whole tightly controlled by the local authorities in the area.

However as the structural constraint to policy making had been lifted the

Figure 6.3 Decision-making Frameworks

Framework	Tyne & Wear Local Authority	Tyne & Wear Urban Development Corporation
Desired planning framework	Strategic	Strategic proactive
Desired internal organisation	Departmental (large)	Small teams
Desired personnel practice	In-house and consultancies (private sector)	Local authority offices, private sector, consultancies and local agencies
Structure of competition	Inter-district competition. Agnostic though generally co-operate with private sector and with central government	Consensus sought with all districts though competitive with one (N Tyneside) district. Competitive with central government. Limited conflict over industrial strategy
Structure of co-ordination and collaboration	Inter-departmental, Intersectoral Interagency	Inter-organisational Intersectoral (Sub regional)
Desired interagency interface	Strategic Hierarchical	Strategic hierarchical fluid, as and when meetings

approach in the UDC progressed some of the trends developed by the local authority and altered direction. For instance the dominance of the rational policy model faded as trial and error, risk taking and speculative policy measures grew in significance. This was reflected also in the ascent of brokering, entrepreneurial and negotiating skills in the UDC. It has to be said however that these were all present in the strategies and the processes of the local authority, but were tempered by the nature of the democratic process and by the culture of a multi-departmental and thus multi-responsive organisation.

Figure 6.4 Decision-making Styles

Style	Tyne & Wear Local Authority	Tyne & Wear Urban Development Corporation
Dominant value	Public good	Economic and public good
Dominant ethic	Public professionalism and entrepreneurialism	Entrepreneurial
Dominant skills	Managerial opportunist	Brokerage and facilitation
Attitude to competition	Within local authority and between districts	Inter-regional
Approach to development	Co-ordinated and planned approach to development	Co-ordinated and planned approach to development
Rules of engagement	Rules clear about democratic action and participation	After initial period rules of engagement developed, participation contingent on access granted by UDC
Legitimation of economic development planning	Legitimation of economic planning decisions through state form and democratic mandate	Legitimation through local actors and communities at local level/3 'E's and project success at central level
Approach to opposition	Co-option of opposition and stabilisation and equilibrium	Co-option of opposition. Stabilisation or equilibrium with possibility of exclusion
Approach to consultation	Informative	Informative, participative
Control of implementation mechanisms	Control of provision	Regulation of external groups
Desired decision-making mode	Rational	Pragmatic opportunist

In Tyne and Wear there is a concentration of national finance producing substantial regional benefit from the redistribution expenditure of central government, there is also an emphasis on labour market strategies. However at the same time local community benefit, social housing and participation strategies suggest that TWDC have sought to ameliorate the effects of social polarisation.

Tyne and Wear has pursued brokerage and consensus relationships within and between local authorities ensuring a regional approach to interurban competition. It is thought to be important that local level politicians and community groups are 'on board' Those not on board and those who do not get access include local MPs and members of national left political networks.

Overall there is substantial continuity in the decision-making framework, except that one can detect slightly more flexibility in the UDC. There is also considerable continuity in the decision-making styles, with the exception of the rules of engagement and the legitimation of the economic planning function. Although the UDC has greater discretion and control of its decision-making environment, it uses a style of operation consistent with the stable political paternalism of the previous era.

7 Conclusion

In this book I have suggested that we need to explore further the hypothesis that the UDCs are an experimental ensemble designed to test new state forms within the context of significant changes to the capital accumulation process.The new types of process form which they apparently seek to insinuate within the local state's economic development and regeneration practices seem at first glance to be consistent with a departure from the 'regulatory' framework which characterised Keynesian state policies before 1973. The impetus for their creation can be seen as logically emanating from the crisis in the state element of the 'mode of regulation' which was dominant until the the mid seventies. This crisis precipitated the rise of neo-liberal policies within the Conservative administrations of the 1980's as well as the plethora of neo-Keynesian left experiments within the arena of local government.

But it is evident from the case studies that the impact of this crisis was uneven and that this was due to the configuration of dominant and residual regional economic frameworks in different urban regions across the UK. This precipitated different responses according to class and class fraction positions within those regions, the reflexivity of the actors involved in redistribution of powers and resources within the agencies of regeneration and the effectiveness of a unifying 'concept of control' in compelling the new regional state controlled formations to develop new strategies. These positions were in turn given a spatial dimension as a result of the structured coherence of urban regions. Thus if they are part of a new 'mode of regulation' the development of state processes consistent with that new mode of regulation is progressing at an uneven rate and manifesting different consequences across the national local state nexus in the urban regions that I have examined.

Strengths and Weaknesses of the Research Framework

The research framework that I have used sought to investigate the impact of spatial difference on UDC policy processes. Uneven development may be a well worn term in social science, but it is nonetheless crucial to contemporary epistemologies. I chose to examine this issue through the use of case studies, backed up by a general survey of UDC activity in the UK. The empirical research was underpinned by the theoretical work of Harvey, Overbeek and Offe. Harvey's notion of structured coherence has been a valuable methodological as well as a conceptual tool in approaching the research. Examining the combination of the technological, organisational and socio-political relations in an area has structured the dimensions of the case studies and given a theoretical foundation to the political strategies that are developing in urban locales. The limitation of drawing upon structured coherence however, is that much more detailed work, covering a broader range of organisations and initiatives, would have to be undertaken in order to conclusively prove the relationship between accumulation, coherence and the regional class alliance.

The UDCs, were contrasted with preceding local authorities in their promotion of regeneration strategies and analysed in the context of their inchoate local relationships. I analysed the macro environment of the urban region, and then proceeded to step down a level, to look at the the micro processes of policy making in the local 'governance' arena. I developed a two fold typology of decision-making styles and frameworks, which I used to evaluate the differences between the preceding agencies and the UDCs and UDC implementation structures in different regions. The typologies mapped the dimensions of the policy process within corporations, and also served to measure the type and rate of change taking place in each of the regions which were examined, therefore providing a useful fillip to the concepts of structured coherence and urban class alliances. The typologies proved to be a valuable tool with which to evaluate the notional differences between the Keynesian and emerging post-Keynesian state form. However without additional contextual material, the categories that I have used within the typologies are not sufficiently nuanced to capture the peculiarities and particularities of change. It must also be said that the categories that have been used in the typologies provided more overlapping characteristics than was originally anticipated.

Therefore, it could be said that in trying to bridge the chasm between theoretical discourse and empirical reality or macro-economic processes, the form of the state, and micro processes of decision-making, that I have not quite captured the detail of processes operating at any one level, in sufficient depth. However the approach I have used represents a step forward, even if a somewhat tentative one, in grasping the between economic relations, state relations and the nature of the contemporary policy making process.

Struggles Over Form, Levels of Continuity and Change in State Form?

Drawing on Overbeek, I have argued that the central government position in the transformation was determined by a neo-liberal hegemony allied to the finance fraction of capital. Its 'concept of control' drew upon white collar and petit bourgeois support. But as I have shown in my case studies, local struggle was already taking place to determine the dimensions of power relationships in the restructured urban locales after the 1973 crisis. At this local level, the class forces were also factionalised as a result of the particular dimensions of the local economy and the position of the pre-existing urban class alliance which dominated the state form.

Central government's response to the pressure exerted by crisis and by the changing international economic environment was, amongst other things, to create UDCs. This allowed the selective use of centrally allocated social expenditure for the speculative creation of a built environment consistent with the competitive marketing of place. It necessitated the co-option of new partners into the decision-making process, who are capable of such speculative activity and the establishment of new rules and procedures that can be legitimated through the success of their competitive strategies.

The reason why the results of the strategy have been so variable and have manifested themselves in different ways is because the strategy of unifying groups around the hegemonic ideas is very much dependent on the outcome of preceding struggles. The framework of the struggle that takes place in each region, and the dimensions of that struggle, are shaped by the pre-existing alliances, and I note in this context the lack of a cross-class alliance which was illustrated in the Bristol case study; effectively there was a free for all, which is as it has always been in Bristol. The lack of an alliance also allowed the development of the interesting urban left experiments in the post-1973 period, but equally the lack of a strong public sector galvanising an alliance has meant that this particular grouping is weak without a democratic framework.

It would seem then that autonomous political spaces exist within the context of state agencies and processes; these spaces are possible even though the state is ultimately bound by the process of capital accumulation. This space allows labour more strength in some areas than in others as different factions dominate in different regions. It also means that struggle will take place at different stages in the process, drawing on different methods to secure access.

This relates back to the position of the alliance as we can see if we contrast the cases of Tyne and Wear and Bristol. In Bristol, it was imperative for oppositional forces to use democratic mechanisms because there were no cross-class alliances. Negotiation over rules was not an option other than through formal democratic processes. Whereas in Tyne and Wear there was extensive negotiation 'behind the scenes' drawing on non-democratic procedures. There was a strong alliance and the local authority was the key unifying factor, as indeed it had been in the earlier period. In Trafford Park there seems to be a new group or alliance developing, emerging from the local authority officer and senior member corps and the different sectoral interests, united by new forms of finance.

The variation in regional responses to the Urban Development Corporations emphasises the inadequacy of categorising conflict over the Urban Development Corporation as simply a dispute between the local and central state. This further reinforces the view that the central local relation dichotomy in state policy needs to be understood with reference to a more general struggle and certainly the dimensions of this struggle do indicate a substantive difference in the strategies that are being pursued in the present era. It is feasible therefore that the struggle is over the imposition of a new mode of 'state regulation'.

There have been a number of disputes about the way that UDCs operate as well as the strategies they pursue. This has manifested itself in different forms throughout the case study regions.

In Bristol where there was least coherence and most fragmentation in class fractions, there was an intense and visible struggle in the pre-designation period. Maximum use of democratic avenues was made by the local authority. After the designation these avenues were closed. Private negotiation took place between the UDC and local authorities, which attempted to secure access to and accountability of the decision-making process under the Sheffield Formula. The inability of the UDC to negotiate in this forum was not really surprising given its association with the dominant finance and property factions of capital. Nor was it surprising that there is no working arrangement between the two organisations as the UDC does not need to negotiate with local authorities. Only in public enquiries were those rights returned to local authorities and the constituents that they represented, but invariably these debates were not about form but about strategy.

In Trafford Park the consensus about how to do things was broken by a neighbouring authority working in consultation with another UDC and other interests in the Park. Even then their disagreement was expressed through negotiations over a strategic issue. Manchester City Council was entitled to object to the UDCs plans for the provision of office space at Wharfside. But in doing so they sought to incorporate demands for member representation in UDC City Council negotiations. The decision to grant this was discretionary and was refused by the Corporation. The power of the actors in this case was dependent on the negotiating skill of the local authority and the forces behind it. TPDC, in the context of an emerging alliance of officers and professionals and a weak property market, gave way on the strategic issue but held their ground on the procedural one.

Conflict in Tyne and Wear also occurred after the designation. Here compromise over the involvement of local communities in the development process around flagship sites was seen by the UDC as a method of quelling local discontent which threatened to break the overall consensus. Negotiating within 'the rules of the game' seems to describe the process in which the leadership and leading officials from local authorities participated. Negotiating about 'the rules of the game' was reserved for community activists, back bench councillors and the political opposition. Success in negotiating with the UDC was more likely when negotiating within the rules, though not impossible outside them, in relation to community demands and in this case it can be suggested that the consensus and strength of the alliance was the dominant factor.

In fact most struggles about the 'rules' were won by UDCs. In securing concessions about process since incorporation of the UDC local authorities have also been singularly unsuccessful. This brings to centre stage the design of the process of policy-making in the UDC.

It is clear that most of the struggles over form have taken place through, or been facilitated by, representative authorities and judicial rights. This gives a certain plausibility to Offe's conclusion that decision rules within the state policy-making processes are often changed when a transformation or crisis takes place in the accumulation process. If, as it seems, the state is altering the rules and here limiting access to the process, it has not happened unopposed. The use of national networks to override coherence and underpin strategies of opposition and infiltrate the process has been visible, but again these have been more successful in securing change on strategic issues such as community benefit. It is clear that the role and embedded nature of the local authority in the alliance is absolutely crucial to any ability to negotiate successfully with a Development Corporation once it is in situ.

Structured Coherence: The Links Between the Economy and the State Form

It has been suggested by Jessop and Harvey that the economic contours of the urban regions contain a mixture of modes of production. Thus we have seen in the case study areas an overlap of dominant and less prominent regimes, which in turn has had an effect upon the interests which are involved in the state policymaking arena. Regions, even within a nation state, exhibit very different structural economic compositions. This suggests the plausibility of conceptualizing differences in the local manifestation of global capitalist circuits through use of the theory of structured coherence. Using such theory with its elements of technical mix, organisational form and social relations, we can interpret city-region differentiation within a global context.

Looking at the three case study regions clear differences are indeed apparent. In the North East it would appear that the effect of the 1973 crisis was not simply a shift from a Fordist to a post-Fordist regime, as the substantial remnants of extensive accumulation regimes, for instance the shipbuilding, steel and coal industries, coexisted with intensive regimes. It is clear, however, that in the present period the region exhibits labour market and consumption characteristics associated with post-Fordist accumulation. These characteristics include technological change, a move to a service economy, flexibility in wage relations, diversification in consumption strategies, and the polarisation of populations.

There is an alliance in which the local authorities were – and are – pivotal players and in which the UDCs now play a full part, which propels changes in strategy at the same time as changing its form. It is a long time since the region has exhibited a strong tendency towards structured coherence. The driving forces in propelling change towards Fordist development, which effectively meant the central state underpinning consumption patterns and public infrastructure, rather than any significant changes in production, have been the local state and organised trades unions. Indeed in this arena of social relations, there has been significant continuity as the technological mix has undergone a transformation around it. In the present period the local state is playing an equally proactive role in grouping together capitalist and previously state monopoly tendencies, but within a new framework. Thus here it would appear that the UDCs are subject to two completely opposing influences, on the one hand they are propelling the 'organisational form' or culture, in a new direction in order to transform the balance in local social relations and the capacity for new technological advances. On the other they retain considerable continuity in the relations around this form, preventing them from undertaking a complete break with the past.

In the North West there is a similar co-existence of intensive and extensive patterns of accumulation, though because of the nature of Trafford Park - an industrial estate built to house factories created by for example Henry Ford and Westinghouse - intensive patterns of Fordist accumulation predominate. It was these industries which were decimated during the crisis of the 1970s and 1980s. Here too there is a move towards the characteristics described by post-Fordists. What is interesting about the alliance here is that the focus is two fold. On the one hand local groupings have developed in the UDA and on the other a conurbation-wide new elite of officers has developed strategic considerations. It appears that a high degree of coherence existed at the point of the transition to Fordism, overlapping somewhat with Keynesian state management. But there have been discontinuities in the alliance. Recently it would appear that there is some movement towards building a new alliance. This is even more important to the locality but difficult because of the mix of economic trends in the locality. Here they did not rely as much as Tyne and Wear on regional policy, rather they relied on Keynesian infrastructural development which underpinned Trafford in particular. Trafford also stood to benefit from the redistributive polices targeted at the Manchester city centre as long as they were complementary to Trafford's policies. Such a contrasting series of local economies in Manchester has made it difficult during the Fordist period to achieve political coherence. Thus while Trafford does not experience the continuity of the alliance in Tyne and Wear, neither does it have to experience the constraint that this would impose on the introduction of new technological strategies. However although Trafford is subject to many of the structural changes in organisational form, it may be hampered in pursuing them fully by the domination of officer skills from Trafford borough.

Finally Bristol has also been experiencing the shake-up of Fordist structures, but interestingly, because of its earlier strategies of diversification and the relative lack of an extensive phase of accumulation, the movement into post-Fordist regimes is advanced. Technological mix dominates the Bristol structure and there is no clear alliance here, but rather a series of significant fractions in competition. The UDC has become allied to one of these. Here the local public sector has been unable - or unwilling - to get into the position to craft an alliance. The lack of coherence has appeared to spur Bristol's economy into further bouts of competition, independent from all but the infrastructural and defence policies of the state.

Reformulation of the Class Alliances

In effect the Trafford case study illustrates a professional alliance operating a strategic decision-making framework in the subregion and locally specific clusters with a greater mix of public private interest feeding into this. In Bristol the lack of any perceptible alliance and underpinning coherence and the split between the UDCs and the other agencies of government including the regional DOE creates a minimum of interaction between agencies. Where this might have occurred, for instance over the issue of training, it does not because there is no strategic overview in the region, although interestingly this has begun to change in recent months in part as a response to changes in the defence industry and also as a result of the process of developing a City Challenge bid. This means that the UDC is in a competitive position vis a vis the other agencies in the city and its alliances are almost solely in the property and finance sector. In Tyne and Wear there is a strong class alliance as a result of public interest domination over the economic agenda since the inception of regional policy in the 1930s. Although there is a shift in emphasis in Tyne and Wear towards capital rather than sustaining the significant hold of representative of labour.

There is most continuity of the actors at the core of the TWDC, the local authority plays a pivotal role and the coopted arena becomes wider. However there are a good number of new structures, from monitoring panels to the formal non-democratic decision-making structure. There is more central state involvement, less back bench involvement and more active private sector involvement. In Manchester too, there is considerable continuity of the local actors as far as officers of the local authority are concerned. But this really only operates at a very local level. Beyond this a new series of networks is forming, bringing new members to the decision-making arena. But again back-bench member power is diminished and the key actors have become the private sector in the area as well as the professional officers beyond. In Bristol many amenity groups and special interest groups are excluded as well as the local authority actors for the area. There are definitely dual processes operating here with the local authority continuing. Unlike Manchester and Tyne and Wear there is not a great deal of involved central government oversight, but that is because they do not have a markedly different structure.

New Processes

Formally both decision-making styles and frameworks have become more constrained by central government procedures and the regeneration policy field, is quite obviously, more closed to intervention by any other than a select group of actors. In all areas small teams which now constitute regeneration departments (the UDCs) now subcontract to other agencies, including the larger local authorities. Similarly, external contracts are maintained and regulated through control mechanisms of a similar form in all three Development Corporations. It is clear that the notion of efficiency is far more important as a legitimating process for the Corporations in all areas, though it is interesting that there are additional and different sub-legitimating processes in each region. The rules of engagement in the processes of the UDC have become far less discernible to outsiders across all the UDCs. The rules of engagement were much clearer under all the local authorities than they have turned out to be under the UDCs. Though to be fair, the rules are only just beginning to emerge in the UDCs. The UDCs clearly have much more power over access to the rule making arena, and are therefore much more powerful in structuring 'the game' than the local authorities. The internal organisation of all UDCs is the same and all of them desire to keep it so. The control mechanisms show minimal differences across regions, though there are major changes in the Keynesian and post Keynesian frameworks.

However, it would appear that there is also a great deal of discretion operating in local areas under the influence of new agencies. The way that discretion is used seems to depend to a great extent on the power of groups in the locality. Where there is not a strong alliance in an area, it would appear that the democratic framework is extremely important in negotiating over the rules of policy processes. This to some extent is also true in the areas where there is a strong alliance, but here it would appear that inevitably groups already in a commanding position fall back on non-democratic processes of negotiation. The structure of the local economy has to a very great extent already shaped the strengths or the weaknesses of the alliance. The much discussed fragmentation of public sector agencies is taking place, but it is being replaced by a new coordination structure in two of the regions, each having a slightly different style.

It is apparent therefore that the overall framework is not always necessarily the same as the style of decision-making. What does this suggest about the changes which we are seeing? Within the decision-making framework we see a replication of the dominant style of the class alliance where it exists and the rules of the coherence where an alliance is absent.

In Bristol the simple rules of competition dominate the arena, based on the association between technical mix and organisational form predominate without the intermediary of a public sector cross class alliance. Trafford Park interestingly reflects the two-tiered spatial configuration of the alliances in the Greater Manchester region, on the one hand a cohesion between very local interests, on the other competition at strategic level. Tyne and Wear remains entwined in the Keynesian state framework, illustrating the stability of the alliance united around the development of a new coherence, with occasional conflict arising from the embeddedness of the fractions of industrial labour. Industrial capital is largely absent in any local form.

The question of legitimation appears to be both locally contingent and dependent on national political considerations. In all local authorities the democratic mandate was very important, though typically for a predominantly Conservative Council, Trafford was also keenly aware of efficiency arguments. Efficiency and project success have become much more important in the UDC period. Additionally, TWDC works very hard at community legitimation, and Bristol relies on the poor image of preceding agencies to garner legitimation.

The way that opposition is dealt with is very much related to the question of legitimacy. Here there are marked differences between the local authorities and the UDCs. The former tended to favour co-option of opposition, though their success in doing this varied. The UDCs use their potential to exclude more than their capacity to include. Here again TWDC is most committed to inclusive strategies, whereas Bristol particularly, but also Trafford Park UDC, seek to delegitimate oppositional interests (TPUDC has sought cooption more frequently than Bristol). Finally, there is a marked difference between the favoured policy design formula for the Keynesian and post-Keynesian era, rationality is the disposition of the local authority while a more eclectic approach emanates from the UDCs.

The planning framework was in all areas strategic, whereas all UDCs operate differently. TWDC is responsive to the demands that are made upon it, and seeks to maintain, with some exceptions, an overall strategic framework, whereas TPUDC is much more flexible and, as expected, Bristol ignores the strategic framework.

Personnel practice is different. Both TPUDC and TWUDC authorities use a mix of inhouse and external services, but TWDC draws much more on local authorities than Trafford Park, whereas Bristol completely eschews the local authority.

A Model for Evaluating Changing Policy Paradigms

To what extent then can we say that there has been a successful implementation of the Thatcher project within the contours of the local state? Has the Thatcherite 'concept' of control become dominant within the state form? What resistance has it met and what has been the result of the resistance? I have developed the notion of policy frameworks and of policy styles in order to illustrate both the structure and the pattern of state policy making as it evolves around the UDC. I have sought to create an ideal model which is a composite of the generalisations that have been made about typical state forms in a Keynesian and a post Keynesian state. I am as much measuring the actual situation against the ideal typical model as I am measuring the real dimensions of change in the areas. The model therefore necessarily synthesises and concentrates the available information about the political economy of urban regions.

As a model the scheme which I have presented has many different uses, but its primary value is that it is possible to substantiate or repudiate the range of claims made about the nature of the current transition. The model presented here is only a baseline from which we can delve more readily into the nature of struggles over these forms.

The efficacy of the dualism of Fordist/ post Fordist or Keynesian /post Keynesian, has been questioned elsewhere (Jones 1985) and in order to accommodate some of the criticism labelled at such simplistic divides I have further sought to break down my typology of policy styles and frameworks in to a series of policy paradigms.

Figure 7.1
Frameworks for Local Economic Development and Urban Regeneration in the Keynesian and Post-Keynesian State

	Local Authority (Keynesian)			**Urban Development Corporaiton** (Post Keynesian)		
	Bristol	**Trafford Park**	**Tyne & Wear**	**Bristol**	**Trafford Park**	**Tyne & Wear**
Desired planning framework	Strategic	Strategic	Strategic	Non strategic, flexible	Strategic, flexible/responsive	Strategic proactive
Desired internal organisation	Departmental (small)	Departmental (small)	Departmental (large)	Small teams	Small teams	Small teams
Desired personnel practice	In-house	In-house and use of consultancies (private sector)	In-house and consultancies (private sector)	Consultancies	Local authority offices and private sector consultancies	Local authority offices, private sector, consultancies and local agencies
Structure of competition	Competition between local authorities and also central government. Public private sector competition	Consensus with local private sector interests. Competition with other local authority districts rather than with central government	Inter-district competition. Agnostic though generally co-operate with private sector and with central government	Competition between all agencies public and private. Competition with central government	Competition with one district: Manchester, though co-operation with other districts and with local interests in private sector (i.e. Trafford Park competitive with central government)	Consensus sought with all districts though competitive with one (N. Tyneside) district. Competitive with central government. Limited conflict over industrial strategy
Structure of co-ordination and collaboration	Inter-departmental	Inter-departmental Intersectoral Interagency	Inter-departmental Intersectoral Interagency	Within contracts	Inter-organisational/agency Intersectoral (Regional)	Inter-organisational Intersectoral (Sub regional)
Desired interagency interface	Non-hierarchical limited strategic efforts	Hierarchical. Attempts to be strategic	Strategic hierarchical	Fluid ad hoc hierarchical non strategic	Strategic hierarchical but fluid and ad hoc as well	Strategic hierarchical fluid, as and when meetings

Figure 7.2
Decision-Making Styles of Economic Development and Urban Regeneration in the Keynesian and Post Keynesian State

	Local Authority (Keynesian)			**Urban Development Corporaiton** (Post Keynesian)		
	Tyne & Wear	**Trafford Park**	**Bristol**	**Tyne & Wear**	**Trafford Park**	**Bristol**
Dominant value	Public good	Public/economic good	Public good	Economic and public good	Economic and public good	Economic good
Dominant ethic	Public professionalism and entrepreneurialism	Professionalism and entrepreneurialism	Professionalism	Entrepreneurial	Entrepreneurial and public sector professionalism	Private sector professionalism
Dominant skills	Managerial opportunist	Managerial opportunist	Managerial	Brokerage and facilitation	Managerial and brokerage	Opportunist and internal managerial
Attitude to competition	Within local authority and between districts	Within local authority and between districts	Intersectoral, inter-departmental, inter-organisational	Inter-regional	Inter-organisational with occasional inter-regional	Inter-organisational
Approach to development	Co-ordinated and planned approach to development	Speculative and risk taking within a loosely co-ordinated framework of development plan	Co-ordinated and planned	Co-ordinated and planned approach to development	Co-ordinated approach – occasional risk speculation	Speculative approach to development
Rules of engagement	Rules clear about democratic action and participation	Rules clear, engagement, contingent all part of process	Rules clear, participation and action clearly part of democratic framework	After initial period rules of engagement developed, participation contingent on access granted by UDC	Rules clearly spelt out. Minimal access all contingent on permission of UDC	Rules ad hoc, ill determined totally contingent on UDC
Legitimation of economic development planning	Legitimation of economic planning decisions through state form and democratic mandate	Legitimation of economic planning decisions on 3 'E's and measurable primary financial criteria	Legitimation of economic planning decisions through state form and democratic mandate	Legitimation through local actors and communities at local level. 3 'E's and project success at central level	Legitimation via 3 'E's and project success	Legitimation not sought – loosely based on internal ethics and superior to poor performance of existing agencies
Approach to opposition	Co-option of opposition, stabilisation and equilibrium	Co-option of opposition. Stabilisation though strongly limited interests	Co-option of some limited interests though unable to maintain stabilisation	Co-option of opposition. Stabilisation or equilibrium with possibility of exclusion	Delegitimation over exclusion of opposition though limited co-option used	Delegitimise and exclude oppositional interests from process
Approach to consultation	Informative	Informative	Informative, participative	Informative, participative	Post hoc informative	Post hoc informative
Control of implementation mechanisms	Control of provision	Control of provision	Control of provision	Regulation of external groups	Regulation of external groups	Regulation of external groups
Desired decision-making mode	Rational	Rational	Rational	Pragmatic opportunist	Pragmatic rational	Ad hoc

The first is the value paradigm and this incorporates the values, skills and ethics in local authority urban regeneration units and in Urban Development Corporations. Secondly I use an organisational and legislative paradigm which includes the organisational parameters of the planning framework, the internal organisational structure, the personnel practices, the control mechanisms and the preferred interagency interface. Thirdly there is a policy/Strategy Paradigm which concentrates on structures and approaches, the structure of collaboration and co-operation as well as the structure of competition and finally the approach to development. The final paradigm is the rules paradigm which looks at the rules of engagement, the approach to consultation which is used, the approach to opposition and the legitimation of planning and economic development and finally what is the overall decision-making model.

The Value Paradigm

The period associated with the Keynesian welfare state is one where the concept of acting in the 'public good' and in a 'professional' manner dominated the value system of the public sector. As there has been a move towards a neo liberal hegemony with a strong emphasis on cutting public expenditure and lowering the PBSR, the emphasis should have switched towards a focus on economic efficiency.

The only locality in which I observed a transition from the dominant value as moving from the 'public good' to the 'economic good' is in Bristol. Although economic efficiency becomes more of a focus in Tyne and Wear, it retained a strong public service ethic. Apart from the close proximity of the UDC to other public sector agencies, the public sector backgrounds of the executive actors and the long term history of the private sector on public expenditure meant that this value would be 'updated' rather than being replaced. In Trafford Park there was a long authority policy making therefore the 'economic efficiency' question was always an important one. One reason for the straightforward transition in Bristol might be the isolation of the UDC from other public sector agencies and also the lack of penetration of targeted urban programme initiatives aimed at alleviating poverty in the area. Therefore the anti-municipalist culture that dominated Bristol was important in allowing the value of public good to be eroded within institutions.

The second area which I examined in the context of values was dominant ethics, especially among the executive personnel. Entrepreneurialism seems to have become the watchword of the so called post fordist/Keynesian era. Yet in both Trafford Park and in Tyne and Wear there already existed a strong entrepreneurial element to public sector policy making. Again this can be traced to the longevity of regional decline and the general consensus on the strategies of regeneration. This meant that officers in particular were accustomed to putting together funding packages for regeneration and were able to transfer these skills to a new era. They combined this with public sector professionalism which although it became less important in both areas still remain in the present period.

Bristol on the other hand had a local authority latterly committed to professionalism, often to the exclusion of commercial and entrepreneurial attitudes. In fact as the urban left became a key mover in the local authority and within amenity groups, the professionalism of this group often come into direct opposition to the commercial decision makers within the city. Bristol did not suffer the same level of recession at the same time as its more northerly counterparts. The emphasis on professionalism remained in the post UDC period but it was a private sector professionalism induced very much by the backgrounds of officers and members and clearly influenced by the type of project which it was possible to undertake in the Bristol area given that their choices were constrained by (a) political hostility to community development and labour market strategies and (b) conditions that prevailed in the commercial market due to the protracted entry of the UDC into the urban arena and the competitiveness of development sites in neighbouring local authority areas.

In all areas there was an overwhelming emphasis on managerialism as a means of dealing with urban regeneration prior to the imposition of the UDC. This changed in the UDC period. Brokerage became important in both Tyne and Wear and in Trafford Park. In Trafford Park it was combined with a managerial element in Tyne and Wear there was much more whereas emphasis on facilitation. Bristol had to maintain a managerialist form because of the nature of its development project but it was also opportunist.

In terms of the values which inform an organisation it seems that the operating culture of the area is very strong in influencing what happens when there is the impetus to change. The key introduction would appear to be the move towards more acceptable brokerage as a means of operating instead of conforming to a rigid managerial formula. The supremacy of economic good over the public good has failed to materialise (this is only further underlined by a national picture of miserable economic efficiency in the majority of the UDCs) although its incursion into the consciousness of bureaucrats is nevertheless observable. Instead it would appear that it has been bolted on it in areas where there has been a strong level of consensus or cohesion in the public sector or where the public sector is seen as legitimate. Entrepreneurialism is not dependent on the move to a Thatcherite or post-Thatcherite government but instead appears to have more to do with the necessities of economic restructuring which prompted the more northerly authorities into action at an early stage of development and encouraged a diversity of skills in its officers.

Organisational and Legislative Paradigm

Strategic planning (within the overall regional and town planning framework) has been the dominant approach to economic well being since the end of the second world war, and became truly hegemonic in the 1960s. This type of approach was desirable because the expansion of the post war state was about managing growth. Managing decline requires a somewhat different approach. Pioneered in Docklands as being non-strategic this seemed to be the neo-liberal answer to dealing with the impossible expenditure and political implications of economic restructuring. But in fact it was just a different sort of management that was needed.

Within the organisational and legislative paradigm the Planning Framework for urban regeneration has undergone significant change in the national context since the early 1970s, a process hastened and intensified by the onset of the neo-liberal project of Thatcherism. Indeed the UDC itself was one element of this restructuring and the mechanism of regional and sub regional planning also underwent significant change. Prior to the arrival of the UDCs the overall ethos of the planning framework was reasonably co-ordinated and was dominated by a strategic approach.

The transition from this framework to the one conceptualised in the model, a laissez faire, non strategic mode of operation, only occurred in Bristol. Indeed in Tyne and Wear there seems to have been an intensification of the drive to maintain a strategic planning framework. It seems that in Tyne and Wear there were two important factors in maintaining the framework. The first was the impact of the newly proposed Unitary Development Plan and its concurrence with TWUDC plans and the impact of the European Funding regimes in that Tyne and Wear as an Objective 2 area was obliged to be part of the new structural fund regimes which demanded an integrated programme.

There was a strong push in TPUDC to maintain a strategic planning programme but there were more problems for the Corporation in doing so. As a single borough within a large conurbation made up of thirteen local authorities they did not have the collective interest of a number of boroughs to maintain and to some extent the success of some of the UDCs development projects were to be gained at the expense of both the Structure Plan and at least one of the other districts. However like Tyne and Wear, Trafford was part of an objective two area within the context of European structural funds and so needed a co-ordinated framework in order to continue to attract these funds. Therefore it would appear that Trafford Park was more 'flexibly' self interested within a strategic framework, while Tyne and Wear was strategically proactive.

The desired internal organisation underwent a significant and wholesale transformation in all regions during the research period. The desired way of working has shifted from large hierarchical organisations to small flexible teams. Why should this happen in a uniform fashion. It would seem that the work of Hogget (1990) is somewhat illuminating here. He described how flexibility and the creation of a strategic centre and an operational periphery allows maximum control of resources, but it also requires a flexible labour force which assists in the seed bedding of a workforce that is internationally competitive. There has been a shift from large hierarchies to small groups. Hogget has suggested that small multi-functional groups mean that not only are organisational loyalties omitted but professional loyalties are truncated too.

Personnel practice exhibits a similar trend in that there has been a shift from in-house working to contract working. This is most clear in Bristol, where the shift has been from public in-house working to private consultants. In both Tyne and Wear and in Trafford Park, this shift has been mediated. In large part this is because the public sector in those areas are the ones where the skills which the UDCs require are present and in both areas a partnership attitude has long since evolved. Crucially in all areas control lies in the hands of the UDC and the Secretary of State. In the former private sector consultants are used as well as drawing upon local authority planning and engineering personnel in particular, the Corporation is using third sector organisations. In Trafford Park the contracts are with private sector consultants and with the local authority.

Interagency interface is a phenomenon that is increasingly important in the new fragmented environment of governance. In Bristol it is safe to say that there was not really an interface (though attempts were made to create it and City Challenge following the UDC seemed to have the desired effect), I would suggest that this is because there was intense competition between the agencies and seemingly no joint interests. In Tyne and Wear prior to the UDCs arrival there appeared to be a somewhat hierarchical system which was increasingly co-ordinated. This continued and extended in the UDC period, although there are some elements of ad-hocery. Trafford Park was part of a less hierarchical interface in the pre-UDC period. Although there was interagency liaison, it was devised to sort out scraps which the local authority got itself into with other districts. In fact the interface was consolidated in the post-UDC period but only at officer level. In reality, in all areas, there were more agencies operating in and around economic development in the post-UDC period, so the effectiveness of the policies often depended on a tight interface. However there seemed to be more issue based meetings, as opposed to strategic meetings, in both areas. It would seem that the ease with which both UDCs accommodated the interagency dynamic was very much about the way in which pre-UDC styles of operating had been developed.

Control of implementation is one of the few areas where there was no dissimilarities between regional agencies a straightforward transition took place. Control of provision was the norm under the Keynesian state system and regulation was the norm in the post Keynesian system. This suggests that there is a new style of control the delivery as well as the direction of policy.

Policy and Strategy Paradigms

Competition between cities and agencies within cities appears to mark an approach to economic development in the post-Keynesian compared to the Keynesian period. While collaboration and co-ordination between agencies was the hall mark of Keynesian economic planning echoing a corporatist subtext, by the same token it can be associated with organisational stasis and complacency. The injection of competition therefore for central funding and for inward investment would appear to be consistent with the decline of managerial are distributive state strategies and the emergence of a state more concerned with maximising control and minimising resource expenditure.

Structure and approaches to competition showed interesting changes in their configurations. In Bristol competition between agencies and between sectors continued from the Keynesian and post-Keynesian period through to date. But the configuration of competition becomes more complex in Tyne & Wear and in Trafford Park. Both local authorities despite their different political positions sought negotiation over urban regeneration policy with central government. This may stem from the substantial revenue input which central government made into both areas to deal with their economic restructuring through urban and regional policy. TWDC tended to operate in a consensus mode, with some limited competition over the general trends of policy (industrial direction versus service direction) with one particular urban left authority. TUPDC was more competitive generally, but tended to seek a united front with local private sector interests and surrounding local authorities with the exception of Manchester City Council, with whom it fell out over territorial and ideological issues. Here the issue was about development sites as well as about who should be involved in policy making procedures. Both Trafford Park and Tyne & Wear UDCs experienced problems with central government, over financial allocation mechanism and over project development. The competitive styles of the UDCs seems to echo competitive frameworks TWUDC sees itself and indeed has positioned in order to unify a regional economic platform. Trafford Park's attempts to do this were obstructed by internal competition within the region.

When it came to examining mechanisms of co-ordination Bristol remained introspective in co-ordinating urban regeneration post UDC. TWDCs agenda was dominated by strategic and co-ordination concerns as it had responsibility for four local authority areas. TPUDC tended again to search for a co-ordinated strategy but as it was located in one borough only it did not have the same authority as Tyne and Wear and developed a capacity to push out the development boat. Both of these UDCs were experienced in funding applications to the European Union; however the cleavages between local authority areas were sharper in the Manchester area as this area was less homogeneous than Tyne and Wear.

While local authority departmental co-ordination was important in all areas, it was less so in Bristol. The UDCs have less to do in terms of co-ordination because of their management structure. The value that was placed on cohesion in Tyne and Wear remained the same, increased in TPUDC, and remained fragmented in Bristol. This can be attributed to the status of the regional intersectoral relationships and also the importance of supranational funding regimes.

The development framework was in fact subject to both central guidance and intensive local regulation in the Keynesian era. Prior to the arrival of the UDC, Bristol and the Tyne and Wear authorities used a co-ordinated development framework. Both Newcastle and Trafford Park authorities indulged in occasional risk development. Tyne and Wear had the most planned and co-ordinated approach to development. This was because of the unifying position which they occupied. In addition the market was more stable in this area. Where the market was more volatile as in Trafford Park there was a tendency to more risk speculation. Bristol was entirely speculative. It would appear that aside from central legislative concerns about freeing up development the adoption of a particular framework was more about condition of the local property market and general economic conditions.

It seems important that there was an opportunist culture in both Tyne and Wear and in Trafford Park prior to the arrival of the UDC. In both areas there are a large number of agencies and the approach to development seems to reflect the style of the dominant partners in the regional class alliance and its affiliation to a particular type of urban policy regime associated with the decline of strong industrial local economies and strong public sectors.

The role which Bristol Development Corporation has played has had little impact on the processes of economic development. The labour market strategy is being shaped now by the European funding mechanism and the new crisis hitting the Bristol labour market. The relationship between local economic development functions and the other redistribution functions of local government had always been somewhat tenuous, so separating it out had little impact; there has always been a dual system operating in Bristol.

The sort of process developing in Trafford Park seems to have been built on what happened prior to the UDCs arrival. The trends that were evident seem to consolidate and expand. The shift towards a non-democratic arena is very strong as is the emphasis on place improvement rather than marketing as such. In Tyne and Wear the UDC sought to ameliorate the polarisation by establishing a two pronged approach i.e. creating social housing and engaging in community development.

Rules Paradigm

The first area that I looked at was the rules of engagement with the agency. How to engage with a local authority over economic development was much clearer in all areas than they have turned out to be in the UDCs, though to be fair the rules are just emerging in the UDCs. The UDCs clearly have much greater powers of access to rule making and are therefore much more powerful in structuring the game than local authorities. The level of individual discretion in each of the areas has become significant. This is crucial because in the context of the new agencies it means that democratic practices are mediated by spatial configurations to a much greater degree than was previously the case.

It is interesting to note that of the different case study areas, Bristol did not have a coherent cross-sectoral class alliance There were a number of disparate and competing groups operating on an informal basis and always if possible beyond the control of the municipal administrative arena. This allowed the municipal arena to be the site of some interesting urban left experiments during the 1970s and the 1980s, but it means that these efforts were isolated from the rest of the decision-making arena as it affected Bristol. Once this arena was closed down by the action of central government, it was very much more difficult for a public sector agenda to achieve substantive results. Therefore here intervention in the decision-making process and discussion and struggle around the decision-making process took place at certain states within the development of the UDC. The ability to fight against the processual idea propounded by the UDC was really only possible before the designation of the UDC. Therefore the nature of the democratic process legitimised through the taking of the case through the courts. After designation the ability to negotiate behind the scenes in order to shape the rules, to attempt to secure more access and accountability by for instance using the Sheffield Formula was an impossibility. Access was at the total discretion of the UDC and the values and methods of those within the UDC predominated. So what were and who participated in the struggles.

In Tyne and Wear there was much more negotiation over the rules behind the scenes of the level of interaction between the UDC and the local authorities was much more inclusive. To a much greater extent the UDC saw itself as reliant on the local authorities. The public sector anyhow had much more legitimacy in the North East, it was a much more homogeneous area and there had been attempts to get a strategy going that was across sectors. Where the difficulties arose was where there had been an urban left authority backed up by a vociferous community and the dispute although located in the discourse about democracy, was about the type of strategy, industrial versus consumption strategy, though there was one dispute here about the nature of power and how power was used in the locality, this involved a dispute over the action of the development corporation in compulsory purchase. There is a much clearer delineation of the actors who could legitimately participate, i.e. leaders, front benchers and community action involved in post industrial or feminist community action those who were out included MPs back bench councillors and industrial type community activists.

In Trafford Park there was an interesting configuration around struggle. Here it tended to take place in a two tier fashion. Within the Park and the UDC area itself there was quite a lot of behind the scenes negotiations, but the agenda was very much set by the UDC. However, there had been significant interest involved in designing rules had happened in the partnership which had been established prior to the UDC. There was struggle about the territorial issues and the democratic issue where there was an urban left council over Wharfside. The configuration of actors was similar in the areas, but there was an emphasis on excluding members.

The second area was the approach to consultation is moving in three different directions in the UDC arena. Only in Tyne and Wear did I perceive an important participative strand in the UDC. The other UDCs tended to be post hoc informative, though the underlying attitude of the Corporations differ with Trafford Park being more committed to involving people.

Most of the struggle over rules had been in fact won by the UDCs. Local authorities and community groups have been singularly unsuccessful in securing gains on the rules of the game around development corporations. New parameters for the rules around economic development activity seem to have become common. Access is not a democratic right, it has to be earned and more importantly you are unlikely to win it if you are outside the 'chosen' groups, or if you disagree with their objectives.

The approaches to opposition also differed. This is very much related to the question of legitimacy. There have been marked differences here between the UDCs. Only in Tyne and Wear is there a vestige of the previous era. Delegitimisation and exclusion have remained the dominant responses. UDCs use their potential to exclude rather than their capacity to include.

Legitimation of planning and economic development tended to be both locally contingent and dependent on national party political considerations. In all local authorities the democratic mandate was very important. Efficiency and effectiveness was keenly endorsed in Trafford, more so than in the other areas, through Tyne and Wear attempted community legitimation. The desired decision-making styles all move from a position of condoning rationality to having a more eclectic and responsive style.

Final Words and Further Research

Are we moving to a new form of state regulation? Is it consistent with contemporaious changes in the economy? Is there any evidence that there is a definite transition from a Keynesian Welfare state to a post-Keynesian authoritarian competition state? What part has the UDC played in this? This book has not given us answers to all of these questions. It has suggested that in Massey's immortal words, 'place matters' (Massey 1984). The demeanour of the local alliances seems to have had some effect on the progression of Urban Development Corporation policy. In part this explains the regional variation in the implementation of that policy, and the different types of struggles that have emerged over the policies. It would also appear however that central government has displayed considerable strength in unifying actors around a comprehensive concept of control, and that the UDC has been an important element in promoting this. Furthermore it does appear that there is a significant transformation taking place in the styles and frameworks of regeneration policy, not withstanding regional differences. It also appears that there are significant
changes in the way that the state operates in the present period and that perhaps this is only the beginning of the process of redefinition of a liberal democratic state.

But by its very nature the development of the Keynesian welfare state pushed forward the parameters of democracy to such an extent that the reversal of civic involvement in the running of the local state would be a hard task to complete. The UDCs may or may not have been a success in terms of regeneration policy; the more important question is how far they have changed the rules of operation in the British state.

Bibliography

Action for Cities (1988) DOE, London.

Adcock, B. (1984) 'Regenerating Merseyside Docklands: The Merseyside Development Corporation 1991-4'. Town Planning Review 65(3).

Aglietta, M. (1979) *A Theory of Capitalist Regulation*, London, Verso.

Ambrose, P. (1986) *Whatever Happened to Planning*, Methuen, London.

Amin, A. and Robins, K. (1990) 'The Re-emergence of Flexible Economies? The Mythical Geography of Flexible Accumulation Environment and Planning', *Society and Space*, Vol.8, pp 7-34.

Amin, A. (1994) *Post Fordism A Reader*, Blackwell, Oxford.

Babbie, C. (1992) *The Practice of Social Research*, Belmont, Waddington Press.

Bachrach, P. and Baratz, M. (1963) 'Decisions and Non Decisions: An Analytical Framework', *American Political Science Review*, 57.

Barnekov, T., Boyle, R., and Rich, D. (1989) *Privatism and Urban Policy in Britain and the United States*, Oxford University Press, Oxford.

Barrett, S. and Fudge, C. (1981) *Policy and Action Essays on the Implementation of Public Policy*, Methuen, London.

Batley, R. (1989) 'London Docklands: An analysis of power relations between the UDCs and Local Government,' *Public Administration*, Vol 67, June, pp. 167-187.

Beck, U., Giddens, A., and Lash, S. (1994) *Reflexive Modernisation*. Politics, tradition and aesthetics in the Modern Social Order, Polity, Cambridge.

Benwell CDP. (1978) The Making of a Ruling Class. Benwell CDP Newcastle.

Bennet, R.R. and Krebs, G. (1991) *Local Economic Development: Public/Private Initiation in Britain and Germany*, Belhaven, London and N.Y.

Benson, J. (1983) 'Interorganisational Networks and Policy Sectors' in D. Rodgers and D. Whetter (eds) *Interorganisational Co-ordination*, Iowa State University Press.

Benson, J. 'The Interorganisational Network as Political Economy', *Administrative Science Quarterly,* June, pp 229-249.

Blunket, D. and Jackson, K. (1987) *Democracy in Crisis: the Town Halls respond*, Hogarth Press, London.

Boddy, M. and Fudge, C., (eds). (1984) *Local Socialism, Labour Councils and New Left Alternatives*, Macmillan, London.

Boddy, M., Lovering, J., Basset, (1986) *Sunbelt City: A Study of Economic Change in Britain's M4 Growth Corridor*. Clarendon Press, Oxford.

Brindley, T., Rydin, Y. and Stoker, G. (1989) *Remaking Planning: The Politics of Urban Change in the Thatcher Years*, Unwin Hyman, London

Bristol City Council, April 1991. *An Evaluation of Bristol Development Corporation: Two Years On*, Report of the Planning Department, Bristol.

Bristol City Council Research Report (1988) *The Role of Bristol City Council and Avon City Council in Economic Development.*

Bristol City Council, Report of the Management Team, *The Proposed Urban Development Corporation.* February 1988.

Bristol Development Corporation, *Realising the Potential,* ECOTEC 1988, Designation Study.

Browhill, S. (1990) *Developing London's Docklands: Another Great Planning Disaster*, Paul Chapman Publishing, London.

Buchanan, D., Boddy, D. and McCalman, J, (1988) *Getting In Getting On Getting Out and Getting Back In Again* in A. Bryman (ed) Doing Research in Organisations, Routledge, London.

Burgess (1984) *In the Field: An Introduction to Field Research*, Unwin Hyman, London.

Burton, P. Unpublished thesis, *Planning Theory and Public Policy: An Analysis of Policies for London's Docklands.* SAUS 1986.

Byrne, D. (1993) Property Developments and Petty Markets versus Maritime Industrialism: Past Present and Future' in R.Imre and H.Thomas, *British Urban Policy and the Urban Development Corporations.* (1993) Paul Chapman Press. London.

Campbell, B. (1993) Goliath. Britains Dangerous Places, Methuen, London.

Castells, M. (1979) *City Class and Power*, Macmillan, London

Cawson, A. (1982) *Corporatism in Welfare: Social Policy and State Intervention in Britain,* Heinemann, London.

Cawson, A. and Saunders, P. (1983) 'Corporatism, Competitive Politics and

Class Struggle' in W. Grant (ed) *The Political Economy of Corporations.* Macmillan, London.

Central Manchester Development Corporation (1988). *Designation Study*, ECOTEC.

Centre for Local Economic Strategies (1990). First Year Report of the CLES Monitoring Project on UDCs, CLES, Manchester.

Cerny, P. (1989), The Changing Architecture of Politics. Sage, London.

City of Newcastle upon Tyne, Economic Development Committee, *Economic Development Policy and Programme, 1983-85*.

Cloke, P. (ed) (1992) *Policy and change in Thatchers Britain*, Pergamon Press, Oxford.

Cook, P. (ed) (1989) *Localities: The Changing Face of Urban Britain*, Unwin Hyman, London.

Cockburn, C. (1977) *The Local State,* Pluto, London.

Coulson, A. (1989) *The Strategy and Impact of The Teesside Development Corporation,* INLOGOV, Birmingham.

Coulson, A. (1990), 'Flagships and Flaws assessing the UDC Decade', *Town and Country Planning* November, and also in his report *Strategy and Impact of the Teesside Development Corporation*, INLOGOV (1989).

CSE London Working Group (1980) *Alternative Economic Strategy: A Labour Movement Response to the Crisis.* C.S.E., London.

Cullingworth, J. B. (1988) 10th edition. *Town and Country Planning in Britain.* Unwin Hyman, London.

Dahl, R. (1961) *Who Governs.* Yale University Press, New Haven, Connecticut.

Davies, J. (1972) *The Evangelistic Bureaucrat,* Tavistock, London.

Dearlove, C. (1979) *The Reorganisation of British Local Government.* Cambridge University Press, Cambridge.

Department of the Environment, Annual Report, 1993.

DOE Inner City Programmes. A Report on Developments and Achievements. HMSO, London.

Department of the Environment (1992) City Challenge Guidance.

Docklands Consultative Committee (1990) *Ten Years On*, London.

Dror, Y. (1968) *Public Policy Re-examined.* Chandler, San Francisco.

Duncan, S. and Goodwin, M. (1988) *The Local State and Uneven Development*, Polity Press, Cambridge.

Dunford, M. and Kafkalas, G. (1992) *Cities and Regions in the New Europe*, Belhaven Press, London.

Dunleavy, P. (1986) 'Explaining the Privatisation Boom: Public Choices Versus Radical Approaches', *Public Administration*, Vol.64, Spring, 13-34.

Dunleavy, P. (1992) *Democracy, Bureaucracy and Public Choice*, Harvester Wheatsheaf.

Dunleavy, P. and O'Leary, B. (1987) *Theories of the State: the Politics of Liberal Democracy*, Macmillan, London.

Egdell, S. (1993) *Class*, Routledge, New York.

Elmore, R. (1985) 'Forward and Backward Mapping in Policy Implementation' in Federal and Unitary Systems, Hanf, K., and Toonen, T., Dordecht: Martinus Nijhoff.

Employment Select Committee Third Report *The Employment Effects of Urban Development Corporations*, HOC 327-1. 1987-88.

Etzioni, A. (1967) 'Mixed Scanning: A Third Approach', *Public Administration Review*, 27.

Fainstein, S. (1993) '*Urban Redevelopment and Public Policy in London and New York*', Presented to Challenges in Urban Management Seminar, Newcastle upon Tyne, March 25-27.

Foglesong, R.E. (1986) *Planning the Capitalist City: The Colonial Era to the 1920s*, Princeton University Press, New Jersey.

Gamble, Andrew. (1988) *The Free Economy and the Strong State: The Politics of Thatcherism.* Macmillan. London.

Also Duncan, S. and Goodwin, M. (1988) *The Local State and Uneven Development*, Polity, Cambridge.

Giddens, A. (1979) *Central Problems in Social Theory.* Cambridge. Polity.

Giddens, A. (1981) *A Contemporary Critique of Historical Materialism*, London: Macmillan.

Giddens, A. (1984) *The Constitution of Society*, Cambridge: Polity Press.

Gramsci, A. (1971) Selection from Prison Notebooks. (ed) Quinten Hoare and Geofry Noel Smith. Lawrence and Wishart, London.

Gregory, D. and Urry, J. (1985) *Social Relations and Spatial Structures*, Macmillan, London and Basingstoke.

Gyford, J. (1991) 'The Enabling Council - A Third Model', *Local Government Studies*, Vol.17, No.1, Jan/Feb.

Gyford, J. (1991) *Citizens, Consumers and Councils: Local Government and the Public*, Macmillan, London.

Gyford, J., Leach, S. and Game, C. (1989) *The Changing Politics of Local Government*, Unwin Hyman, London.

Habermas, J. (1987) *Legitimation Crisis*, Polity, Cambridge.

Hall, S. and Jaques, M. *New Times - The changing face of politics in the 1990s*, Lawrence and Wishart, London.

Ham, C. and Hill, M. (1985) *The Policy Process in the Modern Capitalist State*, Wheatsheaf, Sussex.

Hammersley, M. (1989) *The Dilemma of Qualitative Method: Herbert Blumer and the Chicago Traditions.*

Hammersley, M. (1993) *Social Research Philosophy, Politics and Practice*, London: Open University in Association with Sage.
Harding, A. (1992) 'Property Interests and Urban Growth Coalitions in the UK: a brief encounter' in *Rebuilding the City*, P. Healey et al. Spon. London.
Harvey, D. (1982) *The Limits to Capital*, Oxford: Blackwell.
Harvey, D. (1985b) '*The Geopolitics of Capitalism in Social Relations and Spatial Structures*' (ed) Gregory, D. and Urry, J., Macmillan, p. 144. London.
Harvey, D. (1985a) *The Urbanisation of Capital*, Blackwell, Oxford.
Harvey, D. (1989b) *From Managerialism to Entrepreneurialism: The Transformation of Urban Governance in Late Capitalism.* Paper delivered to the Vega Symposium, Stockholm.
Harvey, D. (1989a) *The Condition of Post Modernity*, Oxford: Basil Blackwell.
Healey, P. (1990) 'Policy Processes in Planning', *Policy and Politics,* Vol 8, number 1, pp.91-103.
Healey, P. (1983) *Local Plans in British Land Use Planning*, Oxford: Pergamon.
Healey, P. and Barrett, S. (1990) Structure and Agency in Land and Property Development: Some Ideas for Research, *Urban Studies,* Vol. 27, No. 1, pp. 89-104.
Healey, P., Davoudi, S., O'Toole, M., Tavasonoglu, S., Usler, P. (1992) *Rebuilding the City: Property led Urban Regeneration*, E & F.N. Spon, London.
Healey, P., Macnamara, P., Elson, M. and Doak, A. 1988) *Landuse Planning and the Mediation of Urban Change*, Cambridge University Press, Cambridge.
Heclo, and Wildavsky, A. (1981) *The Private Government of Public Money* Macmillan, London, 2nd edition.
Held, D. (1984) *Political Theory and the Modern State*, Polity, Cambridge.
Heseltine, M. (1990 2nd ed) *Where there's a Will*, Arrow, London.
Hjern, B. and Porter, D, O. 'Implementation Structures, a new unit of administrative analysis' in Organisational Studies. 2 1981. pp.211-227.
Hogget, P. (1990) Modernisation, Political Strategy and the Welfare State. An Organisational Perspective'. Studies in Decentralisation and Quasi markets. 02. SAUS, Bristol.
Hoggwood, B.V. and Gunn, L. (1984) *Policy Analysis for the Real World*, University Press, Oxford.
Hudson, R. (1989) *Wrecking a Region,* Pion, London.
Hunter, F. (1953) *Community Power Structure.* University of North Carolina Press. Chapel Hill.
Imre, R. and Thomas, H. (1993) *British Urban Policy and the Urban Development Corporations,* Paul Chapman Publishers, London.
Jessop, B. (1982) *The Capitalist State*, Martin Robertson, Oxford.

Jessop, B. (1990) *State Theory Putting Capitalist States In Their Place.* Polity, Cambridge.
Jessop, B. (1992) 'Fordism and Post Fordism: A Critical Reformulation' in Storper, M. and Scott, A. *Pathways to Industrialisation and Regional Development*, Routledge, London.
Jessop, B., Bonnett, K., Bromley, S. and Ling, T. (1988) *Thatcherism*, Polity Press, Cambridge.
Jessop, B (1994) 'Post Fordism and the State' in Amin (ed) Post Fordism a Reader. Blackwell Oxford.
Jones, M. (1995) *Spacial Selectivity of the State? The regulationist Enogma, Economic Development, Local Struggle and Transformation*, Institute of British Geographers, University of Northumbria at Newcastle.
Keane, J. (ed) (1988) *Civil Society and the State*, London: Verso.
King, D.S. (1987) *The New Right: Politics, Markets and Citizenship*, Macmillan, London.
Lash, S. and Urry, J. (1987) *The End of Organised Capitalism*, Blackwell, Oxford.
Law, C. (1988) *The Redevelopment of Manchester Docks,* Discussion Paper No. 33, Department of Geography, Salford University.
Law, C. (1989) *Inner City Policy on the Ground: The Manchester Experience.* Cities 4, 336-346.
Law, C. The Spatial Dimension of Offices in Metropolitan Areas: A Comparison of Birmingham, Glasgow and Manchester. Manchester Geographer N.S.G. 33-41.
Lefebre, H. (1993) *The Production of Space*, Blackwell, Oxford.
Lindblom, C. (1959) 'The Science of Muddling Through'. *Administration Review,* 19.
Lindblom, C. (1965) *The Intelligence of Democracy*, U.S.A., Free Press/Macmillan.
Lindblom, C. (1979) 'Still Muddling Not Yet Through'. *Public Administration Review*, 39.
Lipietz, A.(1987) *Miracles and Mirages.* Verso. London.
Lipietz, A. (1989) *Towards a New Economic Order: Post-Fordism, Ecology and Democracy.* Polity, Cambridge.
Logan, J.R. and Molotch, H. (1987) *Urban Fortunes: The Political Economy of Place*, University of California Press, Berkeley, California.
Logan, J.R. and Swanstrom (1990) *Beyond the City Limits*, Temple University Press, Philadelphia.
Lovering, J. (1983) *The Economic Development of Bristol,* SAUS Project Working Paper 1, SSRC Inner City in Context Research Paper.
Lovering, J. (1990) 'A Perfunctory Sort of Post Fordism: Economic

Restructuring and Labour Market Segmentation in Britain in the 1980s', *Work, Employment and Society Special Issue*, pp 9-28.

Lovering, J. (1991) '*Regional and Urban Labour Markets: Towards a New Mode of Regulation?*', Urban Change and Conflict Conference, Lancaster University, September.

Manchester City Council (1989). The Heart of the Conurbation. Manchester.

Marks, S. '*The Regeneration Game: Perceptions of the Bristol Urban Development Corporation*', August 1988, Masters Dissertation, SAUS.

Marsh, D. and Rhodes, R.A.W., (Ed.) (1992) *Policy Networks in British Government.* Clarendon Press, Oxford.

Marx, K. and Engels, F. (1975) *Manifesto of the Community Party*, Progress Publishers, Moscow.

Marx, K. (1978) The Eighteenth Brumaire of Louis Bonaparte. Foreign Languages Press, Peking.

Massey, D. (1984) *Spatial Divisions of Labour Social Structures and the Geography of Production*, Macmillan, London and Basingstoke.

Massey, D. and Allen, J. (1988) *Uneven Re-Development: Cities and Regions in Transition*, Open University, Milton Keynes.

Massey, D. and Meegan, R. (1985) *Politics and Method: Contrasting Studies in Industrial Geography*, Methuen.

Macregor, S., and Pimlot, B. (1991) Tackling the Inner Cities. The 1980's Reviewed, Prospects for the 1990's. Clarendon, London.

Meyer, M. (1989) 'Local Politics: From Administration to Management', Paper presented to Cardiff Symposium on Regulation, Innovation and Spatial Development, University of Wales, September 13-15.

Meyer, M. (1991) 'Politics in the Post-Fordist City', *Socialist Review*, 1.

Meyer, M. (1993) 'Urban Governance in the Post-Fordist City', Paper presented to Challenges to Urban Management Conference, Newcastle, 25-27 March.

Middlemass, K. (1979) *The Politics of Industrial Society. Deutsche. London.*

Mies, M. (1993) *Towards a Methodology for Feminist Research*, in Hammersly, M.

Miliband, R. (1973) *The State in Capitalist Society: The Analysis of the Western System of Power*, Quartet Books, London.

Molotch, H., and Logan, J. (1976) Urban Fortunes; The Political Economy of Place. University of California Press, California.

Montgomery, J. and Thornley, A. (1990) *Radical Planning Initiatives*, Aldershot: Gower.

Moore, V. and Catchpole, L. (1983) The Local Government Planning and Land Act 1980. in Current Law Statute Reprints, Sweet and Maxwell, London.

National Audit Office (1989) Department of Environment: Urban Development Corporations Report by the Comptroller General HMSO, London. House of

Commons Select Committee on Employment, Third Report, *The Employment Effects of UDCs*, HC 327 I and II HMSO London.

Nevin, B. (1991) *The Black Country Development Corporation and the Sandwell Housing Action Trust.* Faculty of the Built Environment. Research Paper No 7. Birmingham Polytechnic.

Newcastle City Council (1991), Draft Consultation: Newcastle upon Tyne Draft Unitary Development Plan.

Newcastle upon Tyne (1983) Economic Development Committee Policy and Programme.

O'Connor, J. (1973) *The Fiscal Crisis of the State,* New York, St. Martin's Press.

Offe, C. (1975) 'The Theory of the Capitalist State and the Problem of Policy Formation' in Lindburg, L. and Alford, R. (eds) *Stress and Contradiction in Modern Capitalism,* Lexington: D.H. Heath.

Offe, C. (1985) *Disorganised Capitalism: Contemporary Transformations of Work and Politics*, Polity Press, Cambridge.

Olsan M. (1978), *The Logic or Collective Action Public Goods and the Theory of Groups.* Harvard University Press, Cambridge, Mass.

Overbeek, H. (1990) *Global Capitalism and National Decline: The Thatcher Decade in Perspective*, Unwin Hyman, London.

Painter, J. (1991) 'Regulation Theory and Local Government Structures', *Local Government Studies,* November/December.

Painter, J. (1992) 'The Culture of Competition'. *Public Policy and Administration,* Volume 7, number 1 Spring.

Painter, J. *Changing Modes of Regulation, Production and Consumption Relations in the British Local State*, Paper presented to 'Towards a Welfare State', Middlesborough University of Teesside, September 1992.

Parkinson, M. and Evans, J. (1988) *Urban Regeneration and the Development Corporations: Liverpool Style,* CURS Working Paper 2, University of Liverpool.

Piore, M.J. and Sabel, C.F. (1984) *The Second Industrial Divide: Possibilities for Prosperity*, Basic Books, USA.

Pollitt, C. (1990) *Managerialism and the Public Services.* Blackwell. Oxford.

Potter, S. (1988) 'Inheritors of the New Town Legacy'. *The Planner*, November.

Poulantzas, N. (1978) *State Power and Socialism*, Verso, London.

Poulantzas, N. (1975) *Classes in Contemporary Capitalism.* Verso. London.

Price Waterhouse (1987) Designation Study of the Tyne and Wear Development Corporation.

Public Accounts Committee, 'Urban Development Corporations', House of Commons, 17th May 1989, HC 385, Committee of Public Accounts 20th Report.

Punter, J. (1990) *Design Control in Bristol. 1940-1990.* Redcliff, Bristol.

Rhodes, R. (1981) *Control and Power in Central Local Government Relations,* Farnborough, Hants.
Rhodes, R. (1986) *The National World of Local Government.* Allen and Unwin, London.
Rhodes, R. (1988) *Beyond Westminster and Whitehall: The Subcentral Government of Britain,* Unwin Hyman, London.
Richardson, J. and Jordon, A. (1979) *Governing Under Pressure.* Martin Robertson, Oxford.
Riddell, P. (1985) *The Thatcher Government*, Blackwell, Oxford.
Roberts, P and Whitney, O. (1991) *Interagency Co-operation and partnership in Leeds,* Cardiff 1991.
Robinson, F. (1989) *Urban Regeneration Policies in Britain - Who Benefits?,* Discussion Paper 94, Centre for Urban and Regional Studies, December.
Robinson, F. (1990) *The Great North,* BBC North East and Centre for Urban and Regional Studies, February.
Robinson, F., Lawrence, M. and Shaw, K. (1993) *More than Bricks and Mortar, Tyne and Wear and Teesside Development Corporations:* A Mid Term Report, Durham University.
Robson, B. (1988) *Those Inner Cities.* Clarendon Press, Oxford.
Sadler, D. (1990) 'The Social Foundations of Planning and the Power of Capital: Teesside in Historical Context'. *Environment and Planning D: Society and Space,* Vol, 8, pp. 323-338.
Saunders, P. (1981) 'Notes on the specificity of the Local State' in M. Boddy and C. Fudge, *The Local State in Theory and Practice*, WP20. SAUS.
Saunders, P. (1982) 'Why study Central Local relations', *Local Government Studies*, Vol. 8. pp 55-66.
Savage, M. and Ward, A. (1993) *Urban Sociology, Capitalism and Modernity*, Macmillan, London.
Saward, M. (1995) *'Hollowing out the State: The theories and their implications*, Paper to the Hollow Crown Conference, University of Newcastle, September 1995.
Sayer, A. (1984) *Method in Social Science: A Realist Approach.*
Sayer, A. and Morgan, K. (1985) 'A Modern Industry in a Declining Region: Links between Method in Theory and Policy in Politics and Method' in Massey, D. and Meegan, R., *Contrasting Studies in Industrial Geography*, Methuen, London.
Schofield, J. (1993) 'Increasing the Generalisability of Action Research' in Hammersly, M. *Social Research Philosophy Politics and Practice*, Open University, Sage.
Scott, A. (1990) *Ideology and the New Social movements*. Unwin Hyman, London.

Storper,M., and Scott, AJ. (1992) *Pathways to Industrialisation and Regional Development.* Routledge, London.

Simon, H, A. (1947) *Administrative Behaviour.* Freepress. Glencoe.

Simmie, J. (1981) *Power, Property and Corporatism: The Political Sociology of Planning*, Macmillan, London and Basingstoke.

Stoker, G. (1988) *The Politics of Local Government*, Macmillan, London.

Stoker, G. (1989) 'Regulation Theory, Local Government and the Transition from Fordism', in King, D. S. and Pierre, J. *Challenges in Local Government*, Sage, London.

Stoker, G. (1989) 'Urban Development Corporations: A Review', *Regional Studies*, Vol 23, No 2, April.

Stoker, G. (1989) 'Creating a Local Government for a post-Fordist Society: The Thatcherite Project' in J.Stewart and G.Stoker *The Future of Local Government.* Macmillan, London.

Stone, D.A. (1988) *Policy Paradox and Political Reason*, Harper Collins, USA.

Thompson, G. (1990) *The Political Economy of the New Right*, London: Pinter.

Thornley, A. (1991) *Urban Planning under Thatcherism: The Challenge of the Market*, Routledge, London.

Tym, R & Partners (1986) Trafford Park Investment Strategy.

Trafford Park Development Corporation. Annual Report' (1988-95)

Tyne and Wear Development Corporation (1990-1994) Annual Report's and Financial Statements.

Tyne and Wear Development Corporation. (1989)'Strategy for community Development.'

Wainwright, H. (1987) *Labour: A Tale of Two Parties*, Hogarth Press/Chatto & Windus, London.

Wildavsky, A. (1979) *Speaking Truth to Power: The Art and Craft of Policy Analysis*, Little, Brown, Boston.

Wright Mills, C. (1956) *The Power Elite.* Oxford University Press, New York 1956.

Yin, R. (1989) *Case Study Research Design and Methods,* 2nd Edition, Sage, California.

Young, H. (1989) *One of Us.* Macmillan 1989, London.

Index